Future Housing

Edited by Alessandra Battisti

Published in London, United Kingdom

Future Housing
http://dx.doi.org/10.5772/intechopen.104065
Edited by Alessandra Battisti

Assistant to the Editor: Livia Calcagni

Contributors
Marco Avila-Calle, Jorge Toledo-Toledo, Federico Córdova-González, Giovanni Caudo, Isidore C. Ezema, Taofeek A. Suleman, Regina K. Okorigba, Tony Watts, Andreas Koch, Olubunmi Comfort Ade-Ojo, Ayotunde Anthony Babalola, Sasha Tsenkova, Yasna Contreras, Beatriz Seguel, Livia Calcagni, Alessandra Battisti, Alberto Calenzo

First published in London, United Kingdom, 2023 by IntechOpen
IntechOpen is the global imprint of INTECHOPEN LIMITED, registered in England and Wales, registration number: 11086078, 5 Princes Gate Court, London, SW7 2QJ, United Kingdom

British Library Cataloguing-in-Publication Data
A catalogue record for this book is available from the British Library

Additional hard and PDF copies can be obtained from orders@intechopen.com

Future Housing
Edited by Alessandra Battisti
p. cm.

This title is part of the Civil Engineering Book Series, Volume 1
Topic: Construction Engineering
Series Editor: Assed Haddad
Topic Editor: Amin Akhnoukh

Print ISBN 978-1-83768-176-1
Online ISBN 978-1-83768-177-8
eBook (PDF) ISBN 978-1-83768-178-5
ISSN 3029-0287

For EU product safety concerns:
IN TECH d.o.o., Prolaz Marije Krucifikse Kozulić 3, 51000 Rijeka, Croatia,
info@intechopen.com or visit our website at intechopen.com.

IntechOpen Book Series

Civil Engineering

Volume 1

Aims and Scope of the Series

Civil engineering is a traditional field of engineering from which most other branches of engineering have evolved. It comprises traditional sub-areas like transportation, structures, construction, geotechnics, water resources, and building materials. It also encompasses sustainability, risk, environment, and other concepts at its core. Historically, developments in civil engineering included traditional aspects of architecture and urban planning as well as practical applications from the construction industry. Most recently, many elements evolved from other fields of knowledge and topics like simulation, optimization, and decision science have been researched and applied to increase and evolve concepts and applications in this field. Civil engineering has evolved in the last years due to the demands of society in terms of the quality of its products, modern applications, official requirements, and cost and schedule restrictions. This series addresses real-life problems and applications of civil engineering and presents recent, cutting-edge research as well as traditional knowledge along with real-world examples of developments in the field.

Meet the Series Editor

Professor Assed N. Haddad is a Civil Engineer with a degree from the Federal University of Rio de Janeiro (UFRJ) earned in 1986, as well as a Juris Doctor degree from the Fluminense University Center earned in 1993, and a Master's degree in Civil Engineering from the Fluminense Federal University (UFF) obtained in 1992. He completed his Ph.D. in Production Engineering from COPPE / Federal University of Rio de Janeiro in 1996. Professor Haddad's academic pursuits have taken him to postdoctoral stays at the University of Florida, USA in 2006; at the Universitat Politècnica de Catalunya, Spain in 2010; and at the University of New South Wales Sydney, Australia in 2019. Currently, he serves as a Full Professor at the Federal University of Rio de Janeiro. He has held visiting professorships at various institutions including the University of Florida, Universitat Politècnica de Catalunya, Universitat Rovira i Virgili, and Western Sydney University. His research expertise encompasses Civil, Environmental, and Production Engineering, with a primary focus on the following topics: Construction Engineering and Management, Risk Management, and Life Cycle Assessment. He has been the recipient of research grants from the State of Rio de Janeiro, Brazil: CNE FAPERJ from 2019 to 2022 and from 2023 to 2025. Additionally, his research grants obtained from the Brazilian Government CNP since 2012 last to this date. Professor Haddad has been involved in several academic endeavors, being the Guest Editor of the International Journal of Construction Management; MDPI's Sustainability, Energies, and Infrastructures; Associate Editor at Frontiers in Built Environment / Sustainable Design and Construction; Guest Editor at Frontiers in Built Environment / Construction Management; and Academic Editor of the Journal of Engineering, Civil Engineering Section of Hindawi. He is currently a Professor of the Environmental Engineering Program at UFRJ and the Civil Engineering Program at UFF.

Meet the Volume Editor

Alessandra Battisti is a practicing architect and Full Professor of Environmental Design and Technology of Architecture at the School of Architecture, Sapienza University of Rome, Italy, where she oversees the master's programs in Landscape Architecture and Architecture-Urban Regeneration. She is also the director of the level-II master's program in the Valorization and Enhancement of Small Historical Centers. She is on the faculty board and teaching staff of the Ph.D. in Environmental Design and Planning. She is an expert consultant for the European community, the Italian Ministry of Universities and Research, and the Italian Ministry of Culture. She has published more than 200 scientific publications and 14 books. She is the winner of more than thirty international architectural competitions.

Contents

Preface

This book seeks to envision some future scenarios in housing, outlining a series of possible transformations that will affect global housing models in the coming years. With a view to current times, the essays within are not intended to provide predictions on housing, but rather to try to grasp how social attitudes, economic values, and technologies employed are changing. The essays, which differ in slant and content, identify some drivers of change that are increasingly transforming and influencing global housing, such as global population growth, exposure to climate-driven risks, continuous ongoing economic crises, persistent levels of poverty, migration phenomena, exponential increase in the use of digital technology and consequent digital divide, and the urgent demand for more equal spaces by the most fragile populations.

The contributions of this book center around four main topics that are closely related to one another and linked to the aforementioned factors of change.

- Balancing green technologies and circular design with housing development in developing countries. This topic is about the importance and urgency of implementing sustainable and green technologies in the housing sector, particularly in developing countries such as Africa, which, when combined with institutional and political decisions related to circular economy processes, can promote and raise environmental awareness.

- Climate-proof and climate-smart solutions to address the humanitarian challenges of climate change. The open issue of managing climate and humanitarian emergencies involves the study of new forms of housing, such as innovative types of mobile devices, driven by climate change effects like global warming or sea-level rise.

- Exploring the potential of green and digital regeneration for the existing housing building stock. The assumption is the invaluable potential contained in the existing housing heritage and the need to regenerate and re-functionalize it by designing solutions that foster digital and green transition as well as social cohesion.

- Mitigating social inequality in the housing sector. The diminishing affordability and availability of urban housing in Europe is posing an increasing challenge to an economically and demographically diversified population.

These topics define the boundary conditions of the structure of the essays. Each chapter illustrates the future potentials of housing linked to the unavoidable challenges of the present and the opportunities provided by digital technologies and economic management processes that were previously unthinkable.

The book outlines only some possible futures of housing, and each future necessarily implies several institutional and political visions as well as different site-specific values linked to the very concept and purpose of housing. Indeed, the concept of housing differs according to the diverse contexts in terms of economic, climatic, and social terms. For this reason, the essays provide a heterogeneous insight and a non-uniform snapshot of future visions, which do not claim to be exhaustive but are the result of such a broad, complex topic that is not easy to grasp within pre-established schemes.

The editor would like to acknowledge Livia Calcagni for her contribution in co-editing the book and the Department of Planning, Design and Technology of Architecture, Faculty of Architecture, Sapienza University of Rome, in which the exegesis work and conclusions were drawn.

Alessandra Battisti
Department of Planning,
Design and Technology of Architecture,
Sapienza University of Rome,
Rome, Italy

Section 1

Balancing Green Circular Design with Housing Demand in Developing Countries

Chapter 1

Perspective Chapter: Imperative of Nigerian Demographics for Green Housing

Olubunmi Comfort Ade-Ojo and Ayotunde Anthony Babalola

Abstract

Housing is a basic necessity of life. Provision of adequate housing is a perennial problem confronting the nations of the world. The problem is more challenging in the developing countries like Nigeria. Nigeria with a population estimated at over 200million people is plagued with the shortage of quality and adequate housing. The demographics from the National Bureau of Statistics shows the median age of 22.7 years. More than 60% of the population are less than 65 years. The implication is that more houses will be required to meet the need of the younger population with attendant environmental, social and economic problems. Increased consumption of physical and environmental resources is required for the development of more houses. Implementing green housing requirements in housing development ensures that resources are used sustainably in the provision of housing for the populace now and in the future.

Keywords: demographics, green, housing, Nigeria, sustainability

1. Introduction

Nigeria is a country with rapidly growing population with an estimated yearly increase of 3.5% per annum. The country is projected to be the 4th most populous by year 2050 with a population of 410 million. Interestingly, over 60% of her population are below 45 years of age [1]. Unfortunately, Nigeria is one of the countries with great housing challenge. The current housing shortage is estimated at 28million units. The country needs to provide 1.2 units per annum to bridge the housing gap [2]. Another problem with housing provision in Nigeria is the poor quality of the available ones with attendant social, environmental and economic implications. Efforts to ameliorate the housing problem by subsequent governments in the country have minimally achieve the expected result while various resources were consumed in the process. Apart from the high cost of such housing schemes, there were no consideration for sustainability—social, economic and environmental. The changing global climatic condition and increasing high cost of living have further compounded the challenges confronting quality housing provision. Hence Nigeria needs to evolve ways to adequately provide quality and affordable housing for her citizens now and especially the growing population of the younger ones. One of such intervention is the green housing [3, 4].

Green housing is the provision of housing without unnecessary consumption of physical environmental and natural resources thereby leaving a lighter footprint on the environment. Green housing development ensures economical and optimal utilization of resource by minimizing wastes, reduce carbon emission, making use of renewable and natural energy sources, green hosing provision conserves water through the use gray water, water harvesting, reducing the use of clean water through technologies. Green housing improves the indoor environmental quality thereby improving user comfortability and health. With the dwindling national economic resources and consistently growing youthful population, the adoption of green housing becomes imperative for future housing [4].

2. Nigeria demographics

Demography is the study of the quantitative and qualitative aspects of human population. It shows the distribution of people across the continent, countries, and regions. The quantitative aspect of demography includes the population composition, density, distribution, growth and size while the qualitative aspect includes sociological in nature. These are factors such as education quality, diet and nutrition, race, social class, crime, wealth and wellbeing. Nigeria is the most populous country in Africa and also the most populous among the black nations of the world. Globally, Nigeria is among the 10 top countries with the largest population [5]. Most of this are people in the working ages, between age 15 and 64. The median age of the population is 22.7. **Table 1** shows Nigerian population growth from independence till date and the population projection till 2050. The growth in population have resultant impact on the population density. The table shows that the urban population also continues to increase in proportion to the population growth. This in no doubt compounds the problem of housing provision and shortage in the urban centers.

According to the national bureau of statistics [6]. Nigeria's population has consistently been on the rise. The NBS record shows Nigeria population as consisting more of younger population. By implication, there will be need for quality housing for this growing population in the future. This does not discountenance the acute shortage of quality and affordable housing currently bedeviling the country. The housing shortage in the country has also grown geometrically with the population while efforts to reduce the housing gap have not yielded commensurate results. Available data indicates that housing deficits has continued to grow in direct proportion to the population growth (**Table 2**).

2.1 Nigerian demographic implication for green housing

The national population policy recognizes population factors, social and economic development, and environmental issues as irrevocably entwined and are all critical to the achievement of sustainable development in Nigeria. Consequently, this section discusses the implications of Nigeria's demography for future housing.

Human beings depend on the environment to survive. The population of a people occupying a geographical location represents the population density. The consistent increase in population increases the population density of the country. There is increase in the demand for housing with increased consumption of both physical and natural environmental resources to meet this growing demand. Subsequent housing development policies did not consider the degrading impact of housing development

Year	Estimated population	Density (ppl/km^2)	Urban population	% of urban population
1960	45,137,812	50	6,955,737	15.4
1965	50,127,214	55	8,296,555	16.60
1970	55,981,400	61	9,942,297	17.80
1975	63,373,572	70	12,535,293	19.8
1980	73,460,724	81	16,139,321	22.0
1985	83,613,300	92	21,434,269	25.6
1990	95,269,988	105	28,276,132	29.7
1995	108,011,465	119	34,785,092	32.2
2000	122,352,009	134	42,627,440	34.8
2005	138,939,478	153	54,289,212	39.1
2010	158,578,261	174	68,949,828	43.5
2015	181,181,744	199	86,561,390	47.8
2020	206,153,000	226	106,639,000	51.7
2025	233,692,000	257	129,131,000	55.3
2030	264,068,000	290	153,962,000	58.3
2035	297,323,000	327	180,826,000	60.8
2040	333,172,000	366	209,775,000	63.0
2045	371,119,000	408	241,450,000	65.1
2050	410,638,000	451	275,538,000	67.1

Source: [1].

Table 1.
Population growth from 1960 to 2050.

Year	Population	Housing deficit (millions)
1991	104,000,000	7
2007	145,000,000	12
2010	158,578,261	14
2015	181,181,744	17
2019	184,000,000	18
2021	211,447,000	20
2022	216,844,000	28

Source: [7–9].

Table 2.
Growth in housing deficit trend.

on the environment. Increased population density has created unequal distribution of the population. The urban centers are overpopulated with attendant slums, environmental degradation and poor-quality housing. The cities and villages on the other hand are left unattended and under developed. Deforestation and desertification are major environmental problems plague the country. Housing provision is reported to be the largest contributor to land degradation and greenhouse gas emission. Construction waste generation and management are poor and indiscriminate. Secondly, the age distribution of Nigerian demography leaves much for concern. Large percentage of the population are below 65 years of age. The median age is 22.7. This age group are growing and requiring housing. Majority of this young population are millennial. Given the global economic development and the impact of climate change, the need for a sustainable building development is more critical to the survival of this generation. Thus, the need for quality housing increases.

Efforts to provide quality housing dates back to the pre-colonial era [3, 7]. Since then, various housing development policies were promulgated with various setbacks. It is on record that these housing schemes fell short of achieving the expected result at different times. The successive housing development policies did not consider the environmental impact and the need to reduce the foot print of such housing developments. Lighting and ventilation are not priorities of such policies while consideration for regional priorities are lacking. The policies did not consider sustainability issues such as the life cycle cost implication of the buildings, user satisfaction and comfortability, indoor environmental quality. Many of the housing policies use prototype building structures. A prototype used for housing units in the northern part of the country is also used for the southern part with high rainfall and relative humidity. The north of Nigeria is in the Sahara region while the south is tropical in nature. Housing development programs contains provisions for infrastructure facilities but they lack consideration for minimizing energy and water consumption. The excessive population growth creates great demand for energy and water resources. With a large population of growing young people and general reduction in the globally available resources, Nigeria need to begin the implementation of green principles for future housing.

The various housing schemes failed to achieve desired goals due to high cost of construction and high economic impact of such developments. The buildings were beyond the reach of an average Nigerian [8]. Despite the abysmal performance, the various housing schemes; public and private, by organizations or individuals have continued to do untold damage to the environment. Little or no consideration is given to occupational cost to the user and comfortability. The housing sector is said to be the major generator of green-house gases, the major cause of deforestation giving the expanse of land required for construction. Housing provision destroys the natural eco system resulting in various environmental challenges such as flooding, erosion and heat highland. The sector consumes about 48% of the world's resources and energy for construction and maintenance [9–12]. There is no specific figure on housing contribution to carbon emission in Nigeria. However, as signatory to various international conventions such as the Inter-Governmental Panel on climate Change (IPCC), the government plans to reduce its greenhouse gas emissions by 20% by 2030, when compared to "business-as-usual" levels [13]. The various housing schemes did not also consider the economic implication of such houses. A future house must be environmentally, socially and economically sustainable through the adoption of green housing schemes.

The demand for more housing in the urban centers lead to environmental degradation, increased cost of land due to scarcity and consequent increase in building cost. The social implication is that there is more demand for labor force in the building sector in the urban areas. Most of these labors force that cannot afford reasonable housing live in slums and the shanties.in contrast, is the massive reduction of economic activities in the rural areas as there are no meaningful demand for building labor force. The skilled and unskilled construction workers are rendered unemployed. The implication on the economy of the rural areas are less desirable and better imagined.

A green-housing refers to the building practices and process that are in tandem with green building principles. A green building is that which is a building which its construction and subsequent operation has minimum negative impact on the environment [14]. The construction and lifetime operation assure healthiest possible environment, it represents the most efficient and least disruptive use of land, water, energy and resources. Since the enactment of the Brundtland's commission in 1984, the world has been very concerned about achieving sustainable development. Sustainable development according to the commission is the one that meets the need of the present generation without compromising the ability of the future generation to meet their own needs. The green building Process helps in Reducing Energy consumption without sacrificing comfort levels, reduces Waste generation due to recycling/reuse, reduces Pollution and minimizes loads Carbon Footprint. With the reported 50% Energy saving and up to 40% Water savings, green building for Nigeria will help in providing adequate low-income housing units much desired. To ensure compliance and a frame work for sustainable housing development, the world green building council sets out some basic green parameters for a green building. These are referred to as green building requirements. In compliance with these principles, various countries set up their green building assessment or rating tools. These are used as benchmarks to ensure adequate implementation of green building principles. Leadership in Energy and Environmental Design (LEED) is one of such rating tools. The LEED is most widely accepted and used due to its ease of adaptation. The LEED certification system have seven impact factors [12]. Nigeria is yet to have her own certification tools. However, efforts have been made by the government to ensure housing development in the country are sustainable.

2.2 Efforts towards green building development in Nigeria

Efforts towards green building development kick started in 2014 with the registration of the green Building Council of Nigeria (GBCN) with the World Green Building Council (WGBC) as a probationer member [15]. This is in contrast with some African countries like Malaysia. The establishment of the 3rd Malaysian Development plan of 1976–1980 was followed by establishing the National Green Technology Policy (NGTP) and the introduction of the green Building Index (GBI) [16].

Although, Nigeria is prospecting the adoption of the South African Green Star rating tool, [15] believes it would have no significant impact on investment in green building development in Nigeria. According to him, the Green Star does not imply a serious commitment to green building development. Therefore, the use of the Green Star may not sufficiently improve green building development in Nigeria. It should therefore be a temporary measure for Nigeria to develop her own rating tool. Conclusively, efforts towards green building development are at infancy in Nigeria unlike the duo of South Africa and Malaysia [17, 18].

However, the federal government through the Federal Ministry of Power, Steel and Housing in 2016 developed the building Energy Efficiency Guideline (BEEG). The BEEG is in cognizance of the energy situation in Nigeria and the life cycle cost implication on buildings. Efficient energy performance is a key criterion for green building development. The electricity burden on Nigeria is huge and this has been a major albatross to the housing development program. With a large percentage of the population lacking access to constant electricity supply, any step taken towards energy efficient building is of great importance. The establishment of the BEEG is noteworthy since most green building rating tools kick started from the energy efficient point such as the GBI, LEEDv1 and many others [19–21]. The Nigerian BEEG provides regulations and information on the development of efficient residential and office energy costs [22].

Office ventilation accounted for 40–68% of electrical consumption, lighting (13–37%), and office equipment (12–25%) [23]. Energy required for Heating, Ventilation and Air conditioning (HVAC) for residential buildings in Nigeria is 50% compared to only 15% required for lighting. These have great implication for life cycle costs of the building projects while contributing significantly to greenhouse gas effect. Thus, a meaningful housing provision for the teeming populace must meet the energy requirement for a green housing development. The BEEG in conjunction with other green rating systems will enhance the quality of the future housing provision in Nigeria.

Developing the BEEG included a general review of various building assessment tools. The LEED and Green Star-SA were recommended for use and probable adaptation in Nigeria. The LEED has about 317,039 gross square meters of certified and registered building projects in Nigeria [15]. There are over 135 countries using LEED. Some of the LEED certified buildings in Nigeria include the Heritage Place in Ikoyi Lagos, The NOX building in Abuja (Gold rated), P&G Nigeria MDO warehouse, AfDB Nigeria Field Office in Abuja and some others [24]. Reiterating the need for green housing development in Nigeria, [25] identified lack of awareness, lack of enabling policies and legislation to encourage prospective clients and unfavorable economic situation as challenges to green building development in Nigeria. Undoubtedly, the sorry state of the Nigerian housing sector and the demographics requires the implementation of green building practices to achieve appropriate future housing.

2.3 Future housing

The Nigerian population continued to grow geometrically while housing provision is in its trail. Recently, the minister for Power, Steel and Housing, Babatunde Raji Fasola, declared that there was no housing shortage in Nigeria. This he premised on the fact that many urban dwellers have houses in rural areas especially their native towns and villages not adequately in use. Many of the rural dwellers do not have quality housing while most of the urban settlers live in slums and substandard housing. This claim presents some facts about available housing in Nigeria but does not at any rate diminish the fact that Nigerian housing provision at the current pace cannot meet the future housing needs of the Nigerian demographics [7, 26]. Nigeria needs to construct 1.2 million housing units per year to offset the current housing deficit [27]. The goal of the revised national housing policy of year 2006 was to ensure that Nigerians have access to decent, safe and healthy housing accommodation at affordable price. Hence, to achieve qualitative housing provision, Nigeria future housing must consider implementing the green building requirements. These requirements are established as 5 impact factors. The imperative of this requirement to the Nigerian future housing are as follows:

2.3.1 Site design

The design stage forms the foundation for every building project with great impact on cost and project performance. This requirement under LEED is addressed in two ways. First is the requirement for location and transportation. This requirement ensures ease of movement and unhindered access to public infrastructures to reduce carbon emission through vehicular movement. The buildings are close proximity to public infrastructures and facilities [12]. Poor transportation system, infrastructural deficit is a major driver of rural-urban migration. The rural areas are neglected with poor infrastructure and no enforcement of building regulations. The rural dwellers are unsatisfied. Housing intervention program are targeted at the urban centers leaving the less developed towns, cities and villages poorer and less desirable for dwelling. It is very expensive and disturbing for those in the rural areas, towns and cities to access facilities outside their immediate environment. This include good hospitals, schools, market and other social infrastructures. Developers want good return on their investments; hence there is need for government policy drafted to developing quality housing that meets the need of the younger population. Improve the development of the rural areas, redistribute construction economic resources and minimize the pressure for housing development in the urban centers. If the younger population are assured of adequate affordable housing, with the advent of internet services, rural urban migration will be minimized.

The second aspect is sustainable site. This requirement aims to minimize urban sprawl and needless destruction of valuable land, habitat and green space. It discourages inefficient low-density development, encourages higher density urban development, urban re-development and urban renewal, and brownfield development as a means to preserve valuable green space. The future housing provision should preserve key environmental assets through careful examination of each site. It should engage a design and construction process that minimizes site disturbance, values, preserves and actually restores or regenerates valuable habitat, green space and associated eco-systems that are vital to sustaining life [10]. Minimal effort is required for to achieve this requirement in the southern part of the country being a rain forest zone. Contrastingly, the northern part of the country is ravaged by deforestation. Unfortunately, as well, the eastern part suffers lots of land degradation form gully erosion. The land use Act currently in force in Nigeria also poses a challenge to the attainment of this requirement as it hampers access to land. The difficulty in the processing of land titles, certificate of occupancies and related documents are contributory factors to the development of urban sprawls [28–30].

Urban centers like Lagos, Nigeria are over developed. The need for more sites for housing development created the Eko-Atlantic City. Many highbrow housing areas in Lagos were sand filled and reclaimed from the Atlantic Ocean to provide more land for housing development. The incidence of slums, indiscriminate waste disposal with attendant pollution is on the rise as more people desire to live in the urban centers. Urban centers are better regulated with better layouts and improved access to social and economic facilities. However, re-development and urban renewal activities including brownfield development are at its lowest ebb. The geometric growth of the urban centers generates slums and shanties as building development spills unconsciously to adjourning rural communities without adequate provision or government presence.

New buildings are hardly developed in the rural areas due to the lack of interest of the younger generation to stay back while existing ones are left to dilapidate. The principle of sustainable site if incorporated into housing development policy in Nigeria will ensure that future housing is developed to cater for not only the urban dwellers but the rural dwellers also. The provision of modern and affordable housing units in the rural areas will minimize the incidence of urban sprawl and shanties due to lack of reasonable development in the rural areas.

2.3.2 Water quality and conservation (water efficiency)

Reducing water consumption and protecting water quality are key objectives in sustainable buildings. The construction industry is said to be responsible for more than half of carbon emission, water consumption and land fill wastes in the UK with 13% of the raw materials used [10]. According to the report, about 35% of human water use is unsustainable. The percentage will likely increase if climate change worsens, populations increase, aquifers become progressively depleted and supplies become polluted and unsanitary. Humans currently use 40–50% of the globally available freshwater in the approximate proportion of 70% for agriculture, 20% for industry, and 10% for domestic purposes with the total volume increasing progressively. The low-cost houses are poorly serviced and existing water systems are in a deplorable state [31]. Contrastingly, the report on water usage in Nigeria has 69% for agriculture, 10% for industry, and 21% for domestic purposes [32]. Comparatively, Nigeria uses twice the volume of water for domestic purposes, water conservation should be of great concern for future housing.

Provision of portable water is a major challenge to both the rural and urban dwellers in Nigeria. Consequently, the Federal government declared a State of emergency in the water sector. The Nigerian President, Mohamed Buhari noted that access to piped water services which was 32% in 1990 has declined to 7% in 2015 [33, 34]. With global warming, the volume of available water continues to dwindle. Most urban dwellers in Nigeria buy water from vendors while those in highbrow areas live on boreholes. The rural dwellers make use of whatever is available from streams, to hand dug wells and sometimes borehole water from well-meaning neighbors. Motorized wells and boreholes are more prevalent in towns and cities. Every household have a borehole with grave implication for soil stability. Aside the increased cost implication to the home owner in the cities and towns, the rural dwellers do not have access to portable water. Quality water from public facilities are zero to non-existence. The declared state of emergency has no meaningful impact on its availability. Therefore, the requirement for water efficiency needs serious consideration for Nigerian housing provision.

2.3.3 Energy and environment

Energy efficiency leads to important social benefits, such as reducing the energy bills for poor households [35]. Power generation is generally the most expensive feature to add to a building. It is a major challenge for green building development in Nigeria according to literature, 40 million litters of petrol is consumed daily for private generation of electricity [36]. Surprisingly, green building development produces a high-performance building which uses less energy. Hence, its adverse impacts on the environment (air, water, land, natural resources) is minimized through optimized building siting, optimized building design, material selection, and aggressive use of energy conservation measures. The resulting building performance exceeds minimum International Energy Code (IEC) compliance level by 30–40% or more. It maximizes the use of renewable energy and other low impact energy sources. Embodied energy makes up to 30% of the overall life cycle energy consumption of buildings. Effective window placement for day-lighting is also employed to provide more of natural light against the use of artificial lighting during the day [37].

Residential customers sue up 64% of energy generated in Nigeria, 27% commercial with 9% used by industries. In the same vein, 50% of household power consumption is used for Heating, Ventilation and Air Conditioning (HVAC) compared to only 15% required for lighting in Nigeria [38]. Apart from its contribution to high operational cost, this also increases the greenhouse gas emission especially in urban centers. Since there is no record decrease in the rate of global warming across the world, the demand for HVAC will continue to increase. To meet the housing need of growing Nigerian population, the use of green and renewable energy is pertinent for future housing. Any housing development that does not provide for HVAC is not appropriate. It goes to say that 50% of whatever is spent on electricity is used for HVAC. Buildings in the northern part of the country need electricity for heating more at some point in the year while those in the southern pat needs more for ventilation. Energy cost constitute more of the occupational cost. The provision of green energy using renewable energy sources, energy metering and use of high impact opening for ventilation and lighting, both the initial and future running cost of such buildings will be minimized. This will improve the positive impact of the building.

The rural areas are left in darkness for months up to years, the cities towns and urban centers are subject to noise and pollution from generating sets. There are reports of families losing their lives to fumes from generating sets left running while they went to sleep. Others would have suffered life threatening illnesses not so obvious to the naked eyes as a result of pollution from generating sets. Despite the economic challenges in the country, an average home uses 50% of its income on fueling generating set alone. This is an economic loss which is reversible through the use of renewable energy sources such as solar inverter systems. Improved energy demand can be achieved also by the use of energy saving materials [21].

The Nigerian power sector persistently is unable to provide adequate supply of electricity to domestic households and for industrial uses in contrast to its rapidly growing economy and population. Only 45% of Nigerian's population are connected to the power grid. There is power failure 85% of the time with and average supply of 4 hours/day. The obviously poor supply of electricity in the country is attributed to Nigeria's overdependence on oil revenue, lack of legal framework articulating comprehensive strategy promoting green energy and political will to enforce and implement existing laws. The country's huge potential for renewable energy is not fully utilized unlike countries like china, brazil and Germany. The installation of prepaid meters after the privatization exercise gave a boost to energy metering in contrast, to use of renewable energy sources [39].

Poor power supply cripples developmental and economic activities, negatively affecting the economic power and livelihood of the younger population. The resultant effect is the use of electrical power generating set to meet up their electricity needs. This comes at a high cost to the individual, and environmental pollution from burning of fossil fuel used for the generating sets. Nearly every household in the cities and towns own a particular type of generating set or the other. People in the villages now use smaller generating sets to substitute for electricity which may not be available for 3–4 months and to years in some instances. In 2018, the federal government through the Nigeria Energy Policy (NEP) [24] embarked on rural electrification process to provide solar energy systems to 5million houses by 2030 if successful. The NEP is to guarantee adequate, reliable and sustainable supply of energy at an optimum cost. However, NEP was intent on providing access to power and not on environmental sustainability. Therefore, there is not much consideration for renewable energy sources such as biomass, geothermal, hydropower, solar and wind energy. While the country seeks to promote access to qualitative power supply, there should be balance with environmental sustainability.

Apart from poor power supply, another major problem is that 80% of carbon emission in Nigeria is from fossil fuel [40]. Nigeria currently sources very little of its energy from wind and solar. In 2018, around 18% of its electricity came from hydropower which is the largest source of low-carbon energy in Nigeria's power. In 2006, Nigeria produced a "Renewable Energy Master Plan" (REMP). Updated in 2011, the plan seeks to increase the supply of renewable electricity to 23% of the total electricity generation in 2025 and 36% by 2030 [13].

From the foregoing, the need for energy efficient building in the provision of future housing cannot be over emphasized. The gap in the energy need and supply will not provide the quality housing if this requirement for green housing is not incorporated into housing policies in Nigeria.

2.3.4 Indoor Environmental Quality

The Indoor Environmental Quality (IEQ) category such as in LEED standards was created to provide comfort, well-being, and productivity of occupants. The IEQ requirement provides comfort, well-being, and improve the productivity of occupants [41]. The objective of this principle is to provide a healthy, comfortable and productive indoor environment for building occupants and visitors. The building design affords the users the best possible conditions in terms of indoor air quality, ventilation, and thermal comfort, access to natural ventilation and day-lighting and effective control of the acoustical environment. Human beings spend more than 90% of their time indoors while 87% is spent at home [37].

A careful integration of daylight and electrical light sources improves the lighting quality and energy performance of a structure giving a performance luminous environment. A green building provides opportunity for Personal temperature and airflow control over the HVAC system backed. A properly designed green building envelope increases the building's thermal quality [42]. A well-insulated and tightly sealed building envelope reduces moisture problems. Adequate ventilation is thus necessary to eliminate moisture from sources indoors and prevent incidences of sick building syndrome. A poor quality indoor environment impairs cognitive performance, negatively impacts occupant's health, and reduces performance [43, 44]. However, [45] reported low quality of houses and its services in the government estates.

The impact of global climate change in Nigeria include reduced rainfall, shorter period of the raining season, long term increase in temperature in most part of the country; the extreme northeast, extreme northwest and extreme southwest. The average temperatures increased by 1.4–1.9°C. Heat waves from this phenomenon is expected to be on the rise resulting in hot nights. Hot night is where night time temperatures are in the top 10% experienced by a region. Advances in extreme heat particularly is a threat to millions of Nigerians without access to electricity or air conditioning. 92 in every 1000 people in the urban areas and just 14 in every 1000 in rural areas have access to air conditioning. Future housing must provide for quality indoor environment.

2.3.5 Materials and resources

The requirement seeks to minimize the use of non-renewable construction materials and other resources such as energy and water. Maximizes the use of recycled content materials, modern resource efficient engineered materials, and resource efficient composite type structural systems. Maximizes the use of re-usable, renewable, sustainably managed, bio-based materials [46]. One major factor identified for high cost of building construction in Nigeria is massive importation of building materials used. Apart from the fact that some of these materials might not be suitable for the local climatic conditions, many of these contain CO_2 and are very costly. Green building development seeks to reduce embodied energy and CO_2. Nigerian future housing should be developed with consideration for more environmental friendly building materials, less CO_2 and environmentally resilient.

Nigerian construction sector contributes greatly to waste generation and indiscriminate waste disposal. The building process is subject to rework with high volume of waste generated and Nigeria is yet to embrace deconstruction as a practice. Construction waste management is at infancy. There exist various preventive measures to reduce waste thereby conserving resources and minimizing cost. Building construction process go through conventional systems and the various elements are done in segments. There are incidences of the plumber getting to site to tamper with the tiller's work. The building processes are not well integrated to minimize waste of materials and resources. There are no guides for green material specification. Material choices are made based on visual or ego satisfaction. There is no market outlet for supply and demand for recycled materials. Construction wastes are generally dumped and most times, indiscriminately. Domestic wastes are also treated the same way. Indiscriminate waste disposal in urban centers block drains and causes pollution.

Excessive building materials' cost is one major factor identifies for inadequate housing provision quantitatively and qualitatively [8, 47]. Implementing the green building principle will minimize the use of non-renewable construction materials and other resources such as energy and water through efficient engineering, design, planning and construction and effective recycling of construction debris. Maximize the use of recycled content materials, modern resource efficient engineered materials, and resource efficient composite type structural systems wherever possible. Maximize the use of re-usable, renewable, sustainably managed, bio-based materials (Intergovernmental Panel on Climate Change [46]. Up to 35% of the investment in conventional building process goes as waste [12]. Huge amount of financial resources goes down the drain thereby hampering the provision of much needed housing.

The problem of housing shortage, poor quality building and its facilities and environmental degradation due to the increased use and consumption of environmental and natural resources in the provision of affordable housing will be minimized if the principles of Green building development is embraced. The choice of green materials and equipment influences the implementation of sustainable construction practices. However, the lack of knowledge of green technologies is a challenge to the implementation of green strategies and standards [48, 49].

Construction and Demolition Wastes (C&DW) are generated during construction of new building and civil engineering structures, and during renovation of existing buildings and civil engineering structures or demolished. C&DW consist of debris that is generated during construction, renovation and demolition of buildings, roads, and bridges. Construction and demolition waste (C&DW) accounts for half of the solid waste generated in our environment Sustainable waste management processes implemented on housing provision will minimize cost and prevent large volume of waste on landfills [50].

2.3.6 Innovation and regional priority

These are additional requirements in the LEED rating system. Innovative design and construction practices is meant to improve the quality of the building and increase user satisfaction. Implementing green building practices requires innovation there are few LEED accredited professionals in Nigeria since Nigeria is yet to develop its policy for green building development. Design professionals are reluctant to adopt green technology, materials, and resources necessary to reduce the cost of green building development. The requirement for regional priority is to encourage every country to prioritize the immediate environmental, social and economic conditions. Nigeria has a tropical climate. The southernmost part of the country is affected by monsoon rainfall and is characterized by rainforests and mangroves, the country's middle belt has a tropical savannah climate and the most northern part of the country is arid and hot. Most parts of the country have seen a reduction in rainfall. The government estimates a decrease in average temperature across the country [13]. The south, is affected by changes in timing and duration of the raining season while the northern region is experiencing a steep rise in the frequency and duration of drought. This regional variation in climatic condition holds significant for green housing development in Nigeria. The use of prototype buildings across the various climatic zones in the country have not achieved the desired result. Green housing provides that climatic and environmental conditions of each country, region or state is prioritized in design and construction of housing.

3. Conclusions

Nigeria has a large and growing population. A large percentage of this population are young people with the median age at 22.7 years. Historical population growth and the estimated growth till 2050 in addition with the current housing deficit poses a great challenge to housing provision in Nigeria. Subsequent efforts by previous administration have failed to meet expected target of various housing polices. Despite, these polices and subsequent housing development did not consider sustainability issues. The resultant economic, social and environment impacts of the housing development policies were not considered. To meet the housing needs of this teeming population, more physical and natural environmental resources will be consumed. This will further compound the current negative environmental foot prints of housing provision. There will be increased environmental degradation, massive depletion of physical and natural resources, increased construction waste generation with indiscriminate disposal of such, more deforestation and generation of heat highlands and much more. The economic implication of this is enormous. With the huge amount of both public and individual resources committed to housing provision, the life cycle cost implication and social impact of the future housing will be daunting. Hence, efforts towards housing provision should implement green building requirements.

Green housing will minimize the development of urban sprawl, minimize resource consumption and reduce greenhouse gas generation among others. Green housing provides Nigeria the opportunity to minimize the challenges of qualitative housing provision.

Author details

Olubunmi Comfort Ade-Ojo[1*] and Ayotunde Anthony Babalola[2]

1 Federal University of Technology, Akure, Nigeria

2 Federal Polytechnic, Ado-Ekiti, Nigeria

*Address all correspondence to: ade-ojoco@futa.edu.ng

References

[1] Populationof.net. 2022. [Accessed: September 12, 2022]

[2] Statista. 2022. Available from: https://www.statista.com/statistics/1122838/population-of-nigeria

[3] Olotuah AO. Housing development and environmental degeneration in Nigeria. The Built & Human Environment Review. 2010;**3**:42

[4] Ade-Ojo OC. Awareness of the LEED requirements for green housing development among built-environment professionals in Nigeria. Built Environment Project and Asset Management. 2022;**12**(4):521-536. DOI: 10.1108/BEPAM-02-2020-0032

[5] Etebong PC. Demography in Nigeria: Problems and prospects. Biostatics Biometrics Open Access Journal. 2018;**5**(1):555654. DOI: 10.19080/BBOAJ.2018.05.555654

[6] Nigeria Bureau of Statistics. 2022

[7] Emiedafe W. Housing in Nigeria: Why We Desperately Need 17 Million More Homes. Nigeria: Sapient Vendors LTD; 2015

[8] Moore EA. Addressing housing deficit in Nigeria: Issues, challenges and prospects. Central Bank of Nigeria Economic and Financial Review. 2019;**57**(4):201-222

[9] Shen L, Hao JL, Tam VW, Yao H. A checklist for assessing sustainability performance of construction projects. Journal of Civil Engineering and Management. 2007;**13**(4):273-281. DOI: 10.1080/13923730.2007.9636447

[10] Bashir A, Suresh S, Proverbs DG, Gameson R. Barriers towards the sustainable implementation of lean construction in the United Kingdom construction organisations. In: Arcom Doctoral Workshop. 2010. pp. 1-8

[11] Hussin JM, Rahman IA, Memon AH. The way forward in sustainable construction: Issues and challenges. International Journal of Advances in Applied Sciences. 2013;**2**(1):15-24. DOI: 10.11591/ijaas.v2i1.1321

[12] Ade-Ojo CO, Ogunsemi DR. Cost prediction for green affordable housing in Nigeria. In: Abdullateef O, Zalina S, Zhonghua G, editors. Greening Affordable Housing: An Interactive Approach. 1st ed. Florida: CRC Press, Taylor & Francis Group; 2019. pp. 265-281

[13] Dunne D. The carbon brief profile: Nigeria. Country Profiles. 2020

[14] Gupta JK. Buildings Economics. Bhaddal; 2013

[15] Onuoha IJ, Kamarudin N, Aliagha GU, Okeahialam SA, Atilola MI, Atamamen FO. Developing policies and programmes for green buildings: What can Nigeria learn from Malaysia's experience. International Journal of Real Estate Studies. 2017;**11**(2):49-58

[16] Pandey S. MIT-UTM Impact of green building rating systems on the sustainability and efficacy of green buildings: Case analysis of green building index, Malaysia. Malaysia: 2016. [Accessed: August 24, 2019]

[17] Waniko SN. Implication of agricultural and industrial sector linkages on sustainable development in Nigeria. Journal of Sustainable Development in Africa. 2016;**18**(4):67-80

[18] Ejohwomu OA, Oshodi OS. A review of construction management and economics research outputs in Nigeria: Towards a sustainable future. Journal of Construction Project Management and Innovation. 2014;**4**(sup-1):900-905

[19] Howe M. Housing design Award project winner 2011-New Islington. Man

[20] Ghiran A, Mayer A. The Move Toward Net Zero Energy Buildings, Experiences and Lessons from Early Adopters. Institute for Building Efficiency. Available from: http://www.Buildingefficiencyinitiative.org/sites/default/files/legacy/InstituteBE/media/Library/Resources/Existing-Building-Retrofits/Issue-Brief-Moving-to-Net-Zero-Energy-Buildings.pdf. 2012

[21] Bedon C. Recent trends and challenges of energy efficient and sustainable buildings. American Journal of Engineering and Applied Sciences. 2014;**7**(3):292

[22] Arup. Building Energy Efficiency Guideline for Nigeria. Abuja; 2016

[23] Nigerian Energy Support Programme, (NESP). Promoting Clean Energy Investments in Nigeria. 2013

[24] Amasuomo TT, Atanda J, Baird G. Development of a building performance assessment and design tool for residential buildings in Nigeria. Procedia Engineering. 2017;**180**:221-230

[25] Dahiru D, Dania AA, Adejoh A. An investigation into the prospects of green building practice in Nigeria. Journal of Sustainable Development. 2014;**7**(6):158

[26] Olofinji L. An overview of the national housing policy. Real Estate. 2015;**12**. Available from: https://nigeriarealestatehub.com/an-overview-of-the-national-housing-policy [Online]

[27] Ajayi O, Ukanwa E. Nigeria's housing deficit jumps to 28 million, International rights group. Vanguard Newspaper. 2022

[28] Pushkar S, Becker R. Katz, AA methodology for design of environmentally optimal buildings by variable grouping. Building and Environment. 2005. DOI: 10.1016/j.buildenv.2004.09.004

[29] Adedeji YMD, Olotuah AO. An Evaluation of Accessibility of Low-Income Earners to Housing Finance in Nigeria. 2012;**7**(1):23-31. DOI: 10.5829/idosi.aejsr.2012.7.1.1101

[30] Ibimilua AF, Oyewole IA. Housing policy in Nigeria: An overview. American International Journal of Contemporary Research. 2015;**5**(2):53, 2162-59, 142X

[31] Adetokunbo OI. Housing, neighbourhood quality and quality of life in public housing in Lagos, Nigeria. International Journal for Housing Science. 2012;**36**(4):231-240

[32] Doungmanee P. The nexus of agricultural water use and economic development level. Kasetsart Journal of Social Sciences. 2016;**37**(1):38-45. DOI: 10.1016/j.kjss.2016.01.008

[33] Falaju J. Government Declares State of Emergency in Water, Sanitation and Health Sector. The Guardian; 2018

[34] Afolayan FO. Funding higher education in Nigeria. Journal of Research & Method in Education. 2015;**5**(1):63-68. DOI: 10.9790/7388-05116368

[35] Oyedepo SO. Energy and sustainable development in Nigeria: The way forward Energy. Sustainability and Society. 2012;**2**:15

[36] Africa M. How we made it in Africa. Africa Business Insight. 2017

[37] Simpson JR, McPherson EG, Lichter JM, Rowntree RA. Potential of tree shade for reducing building energy use in the Sacramento valley. Journal of Arboriculture. 1993;**22**(1):10-18

[38] Ijigah EA, Jimoh RA, Aruleba BO, Ade AB. An assessment of environmental impacts of building construction projects. Civil and Environmental Research. 2013

[39] Ezema IC, Olotuah AO, Fagbenle OI. Evaluation of energy use in public housing in Lagos, Nigeria: Prospects for renewable energy sources. International Journal of Renewable Energy Development(IJRED). 2016;**5**(1):15-24. DOI: 10.14710/ijred.5

[40] Ogundipe AA, Okwara C, Ogundipe MO. Co_2 emissions and environmental implications in Nigeria. International Journal of Energy Economics and Policy. 2020;**10**(3):317-324

[41] Lee YS, Guerin DA. Indoor environmental quality related to occupant satisfaction and performance in LEED-certified buildings. Indoor and Built Environment. 2009;**18**(4):293-300. DOI: 10.1177/1420326X09105455

[42] Alhorr Y, Arif M, Katafygiotou M, Mazroei A, Kaushik A, Elsarrag E. Impact of indoor environmental quality on occupant well-being and comfort: A review of the literature. International Journal of Sustainable Built Environment. 2016;**5**(1):1-11. DOI: 10.1016/j.ijsbe.2016.03.006

[43] U.S. Green Building Council. LEED v4 User Guide. 2013

[44] Mihai T, Iordache V. Determining the indoor environment quality for an educational building. Energy Procedia. 2016;**85**(2015):566-574. DOI: 10.1016/j.egypro.2015.12.246

[45] Mohit MA, Iyanda SA. Liveability and low-income housing in Nigeria. Procedia—Social and Behavioral Sciences. 2016;**222**:863-871. DOI: 10.1016/j.sbspro.2016.05.198

[46] IPCC. Summary for Policymakers. In: Solomon, Qin SD, Manning M, Chen Z, Marquis M, Aver KB, editors. Climate Change 2007: The Physical Science Basis. Contribution of Working Group I to the Fourth Assessment Report of the Intergovernmental Panel on Climate Change. United Kingdom and New York, NY, USA; 2007. DOI: 10.1038/446727a

[47] Obi ANI, Ubani O. Dynamics of housing affordability in Nigeria. Civil and Environmental Research. 2014;**6**(3):79-84

[48] Saleh MS, Alalouch C. Towards sustainable construction in Oman: Challenges & Opportunities. 2015;**118**(2015):177-184. DOI: 10.1016/j.proeng.2015.08.416

[49] Dalibi SG, Feng JC, Shuangqin L, Sadiq A, Bello BS, Danja II. Hindrances to green building developments in Nigeria 's built environment: "The project professionals ' perspectives". In: IOP Conference Series: Earth and Environmental Science. Vol. 63(012033). IOP Publishing; 2017. pp. 1-9. DOI: 10.1088/1755-1315/63/1/012033

[50] Omoniyi TE, Akinyemi BA, Nwosu SC. A management approach to construction and demolition wastes in Ibadan Metropolis. International Journal of Scientific & Engineering Research. 2014;**5**(5)

Chapter 2

Perspective Chapter: Promoting Circular Design Strategies in Housing Delivery in Nigeria

Isidore C. Ezema, Taofeek A. Suleman and Regina K. Okorigba

Abstract

Circular economy principles are gradually replacing the linear economy model, which has been found to promote waste and resource inefficiency. The circular model is of particular interest to the built environment due to its benefits in resource optimization and waste minimization. Given the huge housing deficit in Nigeria and the attendant resources needed to mitigate the deficit, circular strategies are apt for the massive housing delivery required to bridge the deficit. This chapter examines the concept of circular economy as it affects the built environment. Specifically, design strategies that tend to promote circular housing delivery are examined. The public housing delivery process in use in Lagos, Nigeria's most urbanized city is evaluated to ascertain its alignment with circular principles. The study found that even though opportunities exist for the massive deployment of circular strategies, its adoption is still very low. The chapter recommends more deliberate actions at the design and implementation stages of housing projects to promote circular economy for the housing sector in urban Nigeria.

Keywords: circular economy, design strategies, housing, Lagos, Nigeria

1. Introduction

Housing is a basic need of human beings and a significant part of the infrastructural requirements of any society. The housing sector constitutes about 38% of the construction industry globally [1]. Over the years, concerns about housing have been directed at availability, adequacy, affordability, and sustainability. Housing availability is a major challenge as housing deficit has been ascribed to some major global challenges, especially population growth and urbanization [1, 2]. Already, more than 50% of the world's population currently lives in urban areas with a future projection of approximately 70% by the year 2050 [3]. This growing population will require functional facilities to sustain livelihood of which housing is one of the most prominent.

Responding to the housing deficit effectively will result in high resource extraction and utilization, which are often associated with negative environmental impacts. Hence, mitigating the impact of housing delivery is important in the efforts to provide adequate housing. Hence, emphasis is shifting toward sustainable options in housing delivery. In this respect, life cycle assessment (LCA) has become a widely accepted methodology for estimating the environmental impact of housing provision toward ensuring sustainability [4]. However, the LCA approach has some limitations, especially with respect to its close affinity with the linear materials and energy flow model as against the circular economy model, which has become a preferred option.

In order to promote sustainability through efficient resource utilization, the circular economy (CE) approach has gained ascendancy. CE refers to an economic growth model that prevents environmental degradation by promoting resource efficiency through waste minimization and adoption of regenerative and restorative practices as against end-of-life approach [5]. The concept is, therefore, closely linked to the sustainable development goals SDGs through Target 12.5, which seeks to substantially reduce waste generation through prevention, reduction, recycling, and reuse (3Rs). The circular economy approach when extended to the housing sector seeks to improve sustainability of housing through the use of circular materials, adoption of circular design strategies, reduction of waste in the housing delivery value chain, and adoption of regenerative strategies in housing design and delivery using innovative processes [2].

In the building and construction industry, the design stage has been recognized as the most efficient and effective stage for adopting sustainable practices in which CE strategies can be explored [6, 7]. According to Fatourou-Sipsi and Symeonidou [8], sustainable building design has become necessary with the enormous environmental impact of the building construction and demolition. It has been estimated that CE will result in 4% of economic growth by 2030 in the EU countries [9]. Hence, the EU through the European Green Deal is active in pushing for a more sustainable Europe [10, 11]. However, developing economies in Sub-Saharan Africa, especially Nigeria, are still grappling with the uptake of sustainable buildings. It is, therefore, necessary to fast-track the uptake, especially in the housing sector through the route of CE adoption.

In Nigeria, housing deficit has been estimated to be up to 20 million [12]. In addition, awareness of circular economy is growing in Nigeria [13]. However, solid waste management practices appear to dominate discourse on circular economy in the Nigerian context [14–16]. Meanwhile, the construction industry is growing with technology adoption remaining rather rudimentary in the housing sector. The construction waste implications associated with low technology adoption in the housing construction sector can be very profound. This presents an opportunity to evaluate the prospects of circular design strategies in the Nigerian housing sector. Hence, relevant literature were deployed to underscore circular strategies applied to the design of housing. Built examples of public housing in Lagos, Nigeria were also evaluated to determine the extent of alignment with circular design and construction principles. Given the push towards sustainable buildings in Nigeria and the gradual ascendance of life cycle assessment studies in the Nigerian built environment, the current paper also makes good effort to align circular design strategies with the LCA framework.

2. Circular economy and the built environment

The primary objective of circular economy is to change materials use from the linear model to the circular model. The linear model is a straight-line flow of materials from extraction of raw materials to product use and ultimate disposal to landfill or through incineration at the end of the product life cycle. The circular model promotes use and reuse of materials, through a process usually referred to as "closing the resource loop." In addition to closing the resource loop, circular strategies can also lead to "narrowing" or "slowing" of resource loops thereby improving the efficiency of processes and extending the lifespans of products [17].

In a linear economy, goods are designed for a single lifetime and disposed of at the end of their useful life (cradle-to-death). In contrast, a circular economy aims to eliminate the concept of waste altogether through continuous use of materials. The idea is that the material from the end of one product's life cycle acts as the input for another product's life cycle (cradle-to-cradle). As a result, demand pressure for virgin material is greatly reduced, thus leading to resource optimization. CE slows down the depletion of natural resources and reduces environmental damage resulting from material extraction and processing of virgin materials [18].

Circular design is at the service of a circular economy. Design is the basis of all innovations in products, services, and systems. Circular design is the application of circular economy principles at the design stage of any product, service, or system. It has been estimated that about 80% of a product's environmental impact is determined by the design process [19]. Hence, designing products for reuse can reduce materials and environmental costs. In addition, it has been estimated that over 70% of a product's life cycle costs and environmental footprint are determined at the design stage [20].

Circular design is of particular interest to the built environment. This is so because the built environment is known to be a heavy consumer of resources and energy, as well as a heavy emitter of carbon dioxide and the attendant consequences to the environment. Specifically, the built environment consumes about 50 percent of global materials resources, 50 percent of energy resources, 40 percent of global water use, and 60 percent of prime land, as well as 70 percent of global timber products [21]. More recent estimates put resource use and waste generation by the built environment at about 40% [22]. In addition, the United Nations Environment Programme [23] estimated that the building and construction sector accounted for 35% of global energy use and 38% of all energy-related carbon dioxide emissions in the year 2019. Hence, the built environment needs strategies to reduce its environmental impact in terms of resource utilization, energy use, and carbon dioxide emissions.

Sustainability has been adopted as a preferred development paradigm to ensure efficient resource use while minimizing waste generation in the built environment. In addition, the metrics deployed in the assessment of sustainability have also evolved and can be grouped into three main categories as enunciated by Forsberg and Von-Malmborg [24]. As a result of these metrics, sustainable buildings have evolved, resulting in the reduction of the environmental impact of buildings and the built environment as a whole. However, sustainable buildings have limitations in the sense that they are based on the linear model, which follows the life cycle path of design, construction, use and disposal [25].

Similarly, uptake of circular buildings can be facilitated through adoption of appropriate metrics. In broad terms, Attia and Al-Obaidy [26] identified four primary criteria for assessing the circularity of buildings namely: carbon footprint of building materials used, reused content of the building materials, disassembly potential and longevity of the building, and building design flexibility and long-time use.

Circularity metrics can be applied at micro-, meso-, and macro-levels. At the micro- or product level, one of the popular metrics for measuring circularity is the material circularity indicator (MCI) as articulated by the Ellen MacArthur Foundation [27]. Other indicators include Material Efficiency Metric, Circular Economy Indicator Prototype, and Circularity Potential Indicator [28]. Meanwhile, Drager et al. [29] referred to six circularity metrics aimed at actualizing the major objectives of circular economy as enunciated by the European Environment Agency.

These objectives of circularity metrics were further summarized into three categories namely: protection of materials stock, protection of the environment, and value retention [30].

One thing that is clear with respect to circularity metrics is the plethora of methods available. It has been observed that this multiplicity of metrics can be conflicting and even confusing, and may sometimes lack clarity [31]. In response to the foregoing, the World Business Council for Sustainable Development (WBCSD) developed a comprehensive indicator-based metric for measuring all aspects of circular economy [32, 33]. The WBCSD framework also referred to as circular transition indicators (CTI) comprises a suite of indicators grouped into three broad categories. The first category (close the loop) measures the effectiveness of closing the material loop, while the second category (optimize the loop) demonstrates how material recovery strategies are optimized. The third category (value the loop) demonstrates the business value derivable from applying circular strategies. The CTI framework aligns very well with the major principles of circular economy as enunciated by EMF as follows: design out waste and pollution, keep products and materials in use, and regenerate natural systems.

Given the differences between linear and circular approaches, it would appear that the metrics are parallel. However, it has been shown that LCA, which is the most scientific metric for linear systems, has some usefulness in circularity metrics. Brandstrom and Saidani [28] indicated that material-based circularity metrics align very well with LCA measures in some specific instances. Also, Saade et al. [34] underscored the complementary roles of LCA and circularity indicators in measuring sustainability, especially in relation to early design of urban projects. Similarly, Weidemann et al. [35] demonstrated the complementary roles of LCA and circularity indicators in measuring sustainability, especially in an industrial production context. Realizing that the closed-loop concept of CE does not always ensure environmental benefits, Mannan and Al-Ghamdi [36] demonstrated that LCA can be beneficial for assessing CE options in product design. Very importantly, Van Stijn et al. [37] proposed and successfully tested an LCA-based CE model for the assessment of circular building products.

3. Circular design strategies in housing delivery

From the literature, circular design strategies for housing delivery can be considered under nine subheadings. Most of the strategies are interrelated and complementary. For example, even though treated separately, design for standardization, prefabrication, and modularization are all related but slightly different concepts in circular design. In fact, prefabrication and modularization should ideally go together [38].

3.1 Design for standardization

Standardization is the repeated production of standard sizes and/or layouts of components or complete structures [39]. Design for standardization is deployed to achieve maximum resource and component recovery at end-of-life to support recycling and reuse. The key considerations to achieve design for standardization are avoiding material off-cuts and limiting the use of varying component sizes and the use of standardized connections between elements [40].

3.2 Design for prefabrication

Prefabrication can be referred to as the off-site production of standardized or customized components or complete structures [39]. Design for prefabrication is also known as design for off-site construction or modern method of construction (MMC). It enhances resource optimization by producing building components or the entire building under strictly controlled conditions. Digital technology, in recent times, has been used to optimize the design for prefabrication in the areas of generative design through building information modeling (BIM), parametric designs, and additive and robotic manufacturing [41]. Some countries, such as Hong Kong, North America, Japan, some regions in Europe, and China, have adopted industrialized housing construction through the embrace of design for prefabrication. Prefabricated housing has been employed in Poland since the 1960s; currently, more than 20% of its population lives in prefabricated housing [42]. In Nigeria, rate of adoption of prefabrication in housing is low even though critical stakeholders are familiar with the advantages that can be derived therefrom [43]. According to Silva [3], prefabrication and modularity in housing construction are cost-effective means of achieving affordable housing and also result in a reduction in environmental footprint, flexibility, and the possibility for future selective demolition and recycling.

3.3 Design for modularity

Design for modularity is deployed to achieve easy assembly and disassembly of building components through lean production of building component modules that cut down on time and are less labor-intensive [40]. Modularization is cost-effective and facilitates timely completion of projects as well as construction waste reduction [44]. Modular construction components are also linked to prefabrication, standardization, and system building. Also, Silva [3] investigated ways to implement circularity in the architectural design process that adopts the design for standardization, prefabrication, and modularity through a research-by-design methodology. This resulted in lesser environmental impact and space efficiency and also tended to support disassembly and recycling. The circular design strategies are summarized subsequently.

3.4 Reversible building design

Design for adaptability and flexibility is also known as reversible building design or transformable building design. Design for reversibility supports multiple resource life cycles by integrating other strategies, which includes the design for adaptation, modularity, standardization, prefabrication, disassembly, up-cyclability and adjustment, and flexibility [45]. According to Durmisevic [46], reversible building design protocol on resource circulation covers three main dimensions: functional (spatial), technical (structure), and esthetic (physical) alongside their associated design indicators. Functional reversibility is a design dimension that involves the change of use of space into another without recourse to further material or component use. Technical reversibility is concerned with the design approach that transforms the whole or parts of a building through the rearrangement of the structural components [45]. Design error has been identified as the major barrier to reversibility.

Design for flexibility can facilitate easy refurbishment in the housing sector to avoid untimely demolition through adaptation and material recirculation potentials at building end-of-Life (EoL), and building lifetime extension [47]. Through reversible building design in the housing sector, it can lead to high-value retention in the environment. A study established that reversible design can lead to a 14% reduction in greenhouse gas emission, which corresponds to 1740 t CO2-eq for building components, such as structural elements, foundation, and ceiling components, as evidenced in the study carried out by Kröhnert [47]. Similarly, Kröhnert [47] investigated the adoption of reversible building design principles in multi-story residence building components, which resulted in a reduction in embodied GHG emission, improvement in re-cyclability and re-usability of components, and retention of environmental value.

3.5 Design for reuse

Design for re-usability involves the direct reuse of elements, components, or the entire building with or no recourse to the introduction of additional resources in new construction. Design for re-usability is facilitated by integrating design for modularity, adaptability, flexibility, standardization, dimensional coordination, building reversibility, and specifying reclaimed materials. This can aid in climate change mitigation [47]. The adoption of expandable housing can facilitate the reuse of building components [48].

3.6 Design for disassembly

Across the literature, design for disassembly/deconstruction (DfD) is the most mentioned literature on circular building design [49]. It is a CDS that focuses on the activities that take place at the end-of-life of buildings to recover building elements and components for reuse thereby minimizing waste. The major impediment to design for disassembly is the use of irreversible connections between elements, and other major considerations for design for disassembly are reduced number of components, lightweight elements, avoiding binders, use of accessible connections, and the use of recyclable and reusable components [50]. Another CDS that works with DfD is the design in layers, and this assists in easy deconstruction. DfD supports adaptability, flexibility, and selective deconstruction. The adoption of design for the disassembly has been recognized to be effective in the management of building components for future reuse, which is to be incorporated in the design of new construction at the early stage [8].

3.7 Use of reclaimed or bio-based materials

This is a design strategy that involves the integration of recycled or reclaimed materials or components or the specification of bio-based or circular materials in a new building wholly or partly to enhance the closing and slowing of the resource cycles. Bio-based materials possess several advantages that dispose of them for use as circular materials [51]. In regions with a large stock of buildings for renovation, bio-based products can be deployed to contribute to circularity [52]. It has also been shown that agro-industrial wastes can be converted to building materials thereby contributing to sustainability and circular economy [53].

3.8 Regenerative and restorative design

The regenerative and restorative concepts refer to two aspects of design that align with nature and promote natural processes. Sometimes, the scope of restorative and regenerative designs can extend to socio-technical systems as a wider context within which buildings are situated [2]. Similarly, they refer to integration of building design with natural and natural support systems such as green and gray infrastructure to achieve a harmonious relationship between building and the natural ecosystem [54]. While restorative designs aim at restoring ecological systems to a healthy state, regenerative designs aim at enabling ecological systems to maintain a healthy state [55, 56]. It, therefore, implies that restoration is ameliorative while regeneration is preventive. In addition, Petrovski et al. [57] investigated the adoption of regenerative design principles in the design and construction of a residential building for refurbishment in Spain through a case study approach. The study revealed reduced costs and minimized environmental impact of the refurbishment process.

3.9 Design for energy efficiency

All energy in their most rudimentary and primary level is derivable from nature even though it exists in various forms. Energy efficiency can be examined both from the embodied and operational dimensions. At the operational level, energy efficiency entails use of low energy-consuming appliances, while at the embodied level, it entails use of low-impact materials for buildings. It can also be considered from the perspective of passive and active strategies. While passive strategies promote natural solutions, active strategies deploy technology to achieve efficiency. Integration of passive strategies and circular design principles has been shown to be complementary to promoting resource efficiency in buildings [58]. In addition, the use of renewable energy is a major component of circular economy as it ensures minimal use of fossil-based fuels and a cleaner energy regime [18, 59].

4. Housing challenge and delivery in Nigeria

Affordable housing has become a major concern in Nigerian urban areas. The urban poor constitutes approximately 50% of the Nigerian population living in urban centers [60, 61]. The urbanization rate is inversely proportional to the quality and quantity of housing in Nigeria with a total population currently estimated at 262 million [62]. According to Refs. [61–63], the Nigerian housing sector has been adjudged as unsustainable due to a myriad of challenges. These challenges have given rise to the formation of informal settlements at both the urban core and fringes [62].

Different approaches and various interventions have been employed by the Nigerian government in housing delivery over the years since the colonial era. They include housing for local workers and expatriates in the face of the bubonic plague of 1928 [64], the postindependence housing programs by the Federal Housing Authority, and subsequently followed by Public-Private Partnership [62, 65]. However, as indicated in **Table 1** and further highlighted by Refs. [64, 67, 68], the housing deficit is spiraling and would require over 50 trillion in local currency terms to fix [69].

As a way forward, Olubi and Aseyan [62] emphasized the need for locally inspired housing designs and construction methods using local materials and techniques in housing delivery to assist affordability. Alabi and Fapohunda [70] also advocated for the adoption of cost-reduction strategies, which can stem from the use of locally available materials, the specification of reclaimed materials, and material optimization through design. The material cost of a building project is the major determinant of the construction cost and poor workmanship during the construction phase result

Year	Housing Deficit	Estimated Population
1991–1993	4–7 million	104 million
2007	8–10 million	145 million
2013–2015	16–17 million	178 million
2017–2019	18–22 million	184 million

Source: [66].

Table 1.
Trend in housing deficit in Nigeria.

in high maintenance cost [70]. Using environmentally friendly construction materials such as timber, compressed earth bricks, lime, hemp, hydra form, stone, cob, and rammed earth will assist in achieving sustainability in the Nigerian housing sector [71]. Also, Okoye et al. [67] pointed out the roles of design strategies in affordable housing delivery in Nigeria and further identified that architectural design influences the affordability and simplicity of core houses.

The adoption and implementation of strategies that align with sustainability and affordability in housing delivery have been heralded to be beneficial to the housing sector globally [71]. Of added importance to the Nigerian housing sector is the adoption of circular strategies. The adoption of circular design for sustainable affordable housing aligns with SDGs 11, 12, and 15 [3, 72, 73]. This is a sustainability dimension that needs to be scaled up in the Nigerian context.

According to Refs. [45, 47], CE is a growing area of research in housing delivery that spans from material to city-scale dimensions. Limited investigations exist on CE in housing delivery [45, 47]. The adoption of circular design in the Nigerian housing delivery system will orchestrate the development of new cutting-edge technologies and economical construction methodologies, which have been found advantageous in the delivery of sustainable and affordable housing for low- and middle-income earners [74]. It was projected that the deployment of economically efficient technologies in housing delivery can lead to a 26.11% reduction in the cost of building [74].

5. The context of Lagos

As the most urbanized state in Nigeria, the housing challenge in Lagos is obvious. The housing deficit in Lagos has been estimated to be about 16% of the total estimated deficit in Nigeria [75]. It has been estimated that Lagos has a housing need of 4.69 million housing units with a housing stock of 1.49 million units, thus leaving a deficit of 3.2 million housing units [75]. Multiple stakeholders are involved in the Lagos housing market, which is dominated by private individuals and organizations. However, government plays the role of policy formulator and regulator.

As part of the social function of government, the Lagos State government has also been actively involved in public housing provision. It has been estimated that between 1999 and 2020, over 7,000 housing units of different typologies have been provided by the Lagos State Government [76]. The first phase of the program was targeted to deliver 3632 housing units in about 13 locations in the city of Lagos (see **Table 2**). Additional units are provided on a continuous basis through direct budgetary allocation and through public-private partnerships. Seven thousand units of

Location of Estate	Number of Blocks	Number of units
Igbogbo	22	264
Sogunro 1	12	144
Sogunro 2	8	96
Shitta	3	36
Igando	41	492
Omole	7	84
Magodo	4	48
Lekki 1 & 2	15	180
Mushin	5	60
Ilupeju	10	120
Sangotedo	45	540
Agbowa	70	560
Ijora-Badia	—	1008
		3632

Source: [77].

Table 2.
Distribution of Lagos HOMS estates phase 1.

housing were projected to be delivered at the end of 2022 under the Lagos HOMS program. Hence, more units are being added as new estates or as extensions to existing ones. However, the units have largely maintained the original design and procurement procedure in the last 10 years.

The key delivery strategies include direct construction of housing, site-and-services schemes, and access to mortgage facilities, among others. Since 2012, the Lagos State Government under the coordination of the Ministry of Housing has been providing housing to residents through the Lagos Home Ownership and Mortgage Scheme (Lagos HOMS). The scheme provides access to both the housing units and the mortgage facilities needed to secure the housing units. The scheme is a multiagency scheme involving key players such as Ministry of Housing, Lagos State Development and Property Corporation (LSDPC), Ministry of Physical Planning and Urban Development, and the New Towns Development Authority (NTDA). The mortgage component is facilitated by the Lagos Building and Investment Company—Mortgage Bankers [78]. Another variant of the scheme is the rent-to-own system where occupants pay rent for a stipulated time after which ownership is transferred to them [79]. About 70% of the public housing provided from 1999 was done under the Lagos HOMS program. Hence, the Lagos HOMS project is used as an index to examine the extent of circularity in the provision of the housing schemes.

From previous empirical studies on the reduction of waste associated with buildings in the study area, the approaches considered most valuable by built environment experts include design for standardization, disassembly, reuse, prefabrication, and modularity [80–82]. These approaches have been referred to as modern methods of construction (MMC). However, there is a well-defined planning and design component that precedes the construction phase. The underlying principle in this respect is resource optimization in terms of time, money, and materials. Most of the materials deployed in modern methods of construction are conventional materials but deployed innovatively. Also, an important aspect of resource optimization literature in the study area is the use of alternative materials such as renewable materials and materials made from byproducts of industrial and agro-based processing [83–85]. In addition, the use of renewable energy and the adoption of passive strategies are considered other avenues for optimizing energy-based resources in the study area [86–90]. From the foregoing, four planks can be isolated upon which the assessment of the selected public housing program can be based namely:

i. adoption of MMC and associated strategies,

ii. use of renewable, bio-based, and waste-based materials,

iii. adoption of renewable energy, and

iv. deployment of passive and regenerative strategies

As a prelude, full description of the building and the procurement process is based on the understanding that the design process is critical in the adoption of circular strategies.

6. The Lagos HOMS project

The predominant typology is a rectangular plan with four floors accommodating twelve (12) residential units of various sizes.

Each floor of the prototype is made up of three apartments (one bedroom, two bedrooms, and three bedrooms). There are three staircases for the combined use of the occupants. Each block, therefore, accommodates mixed-income dwelling units rather than previous government estates that have separate sections for low-income and medium-income dwellers. This prototype has been retained over the past 10 years for subsequent housing units with only minimal modifications (**Figure 1**).

The design intended to incorporate sustainable strategies such as natural ventilation, natural lighting, and use of low-impact appliances such as energy-saving electricity bulbs. However, the prototype design is a very compact design and devoid of the openness required in a warm-humid tropical environment. The compact design (as indicated in **Figure 2**) is effective in optimizing building materials used in the buildings. The highly optimized floor plan has minimal circulation area. In terms of emergencies, only the staircases are the escape route as balconies are nonexistent in the buildings. Rainwater harvesting and storage for use in washing and watering plants was also intended but this was not carried through to the implementation stage. The use of ducts for plumbing pipe installation facilitates easy maintenance.

Natural ventilation and lighting were fairly achieved in the building design. However, cross-ventilation as recommended for a warm-humid tropical environment was not fully achieved. In practical terms, cross-ventilation is deemed to have been achieved when a space has window openings on at least two sides. Specifically, the living rooms of the one-bedroom apartment and the two-bedroom apartment have limited opportunity for cross-ventilation. The atrium introduced in the building is not large enough to encourage substantial air flow. No shading devices were observed

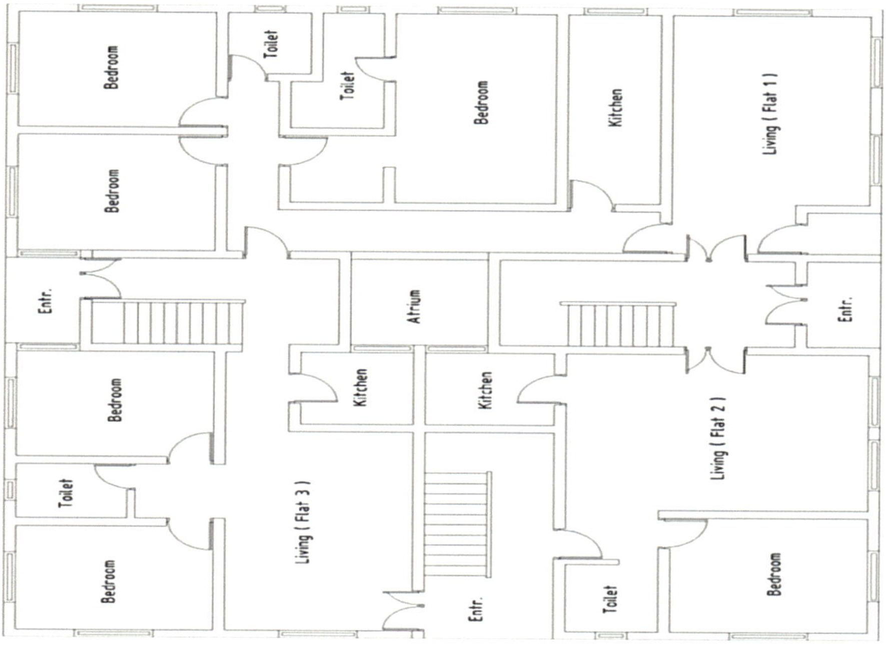

Figure 1.
Schematic layout of a block. Source: Adapted from [91].

Figure 2.
Compact arrangement of the blocks under construction. Source: [91].

in the buildings as built. The lean-to-roof deployed for the project did not have deep overhangs to help shade the walls and openings. Meanwhile, the vegetative cover is low when compared with built-up areas of buildings and paved areas. Hence, passive design principles were not substantially applied (see **Figure 3**). The spacing between buildings is rather narrow relative to building height, which may impair effective air movement around the buildings. Hence, many of the occupants can afford to have opted for air-conditioning for thermal comfort in the buildings.

The construction materials used were conventional materials and the technology adopted was mostly *in situ* construction technology. **Table 3** depicts the materials deployed in the construction of the buildings. The buildings are made of reinforced concrete structural frames (columns, beams, and slabs) with sand-cement blocks

Figure 3.
A Section of the Lagos HOMS layout showing the closeness of the blocks. Source: [92].

Building Component	Main Material Used
Substructure	Concrete, reinforced concrete, sand-cement blocks
Frames and Upper Floors	Reinforced concrete
Walls	Sand-cement blocks (rendered)
Roof Structure	Treated Timber
Roof Covering	Aluminum long-span roofing sheets
Ceiling	Treated Timber noggins, PVC ceiling
Wall Finishes	Sand-cement rendering, emulsion paint, ceramic wall tiles for wet areas
Floor Finishes	Sand-cement backing, Vitrified ceramic tiles
Doors	Steel doors, timber doors
Windows	Aluminum-framed glass
Source: Adapted from [91].	

Table 3.
List of components and materials used for the work.

as external envelopes and internal partition walls. All wall and concrete surfaces are rendered with sand-cement mortar. The roof structure is made of treated timber, while the roof covering is long-span aluminum roofing sheets. The lean-to-roof design optimized the materials used in the roofing when compared to prevailing roof designs in the study context.

Other materials include vitrified ceramic tiles for the floors and glazed wall tiles for the walls of wet areas. The ceiling is made of PVC ceiling tiles supported on timber noggins. External doors are steel doors fabricated and fixed with mortar-to-door openings. Similarly, windows are made of aluminum framed glass fabricated and fixed with mortar to the window openings. Internal doors are made of timber.

The estate road networks are paved with concrete interlocking stones. This helps to manage stormwater drainage within the estates. It also facilitates ease of maintenance without destroying the pavements as paving stones removed for routine maintenance can be reused in the making good process. Vegetative cover for the buildings is limited, which facilitates heat gain in the buildings emanating from the paved surroundings. As a result, residents are resorting to installation of air conditioners for indoor thermal comfort. Electricity supply is from the national grid, while water supply is from dedicated boreholes and associated storage facilities.

6.1 Adoption of MMC and associated strategies

The associated strategies refer to design for standardization, modularization, and prefabrication. They also include design for disassembly and incorporation **of modern construction methods.** The adoption of prototype design options disposes the housing scheme to standardization, modularization, and prefabrication. However, prefabrication was not deployed on a significant scale. Some floor components were prefabricated and installed at some of the Igando Lagos HOMS buildings (See **Figure 4**).

With respect to design for reuse and disassembly, the *in situ* construction adopted for the projects hinders the possibility of disassembly of the building components and their subsequent reuse. The *in situ* reinforced concrete components cannot be reused or disassembled without full demolition. Similarly, *in situ* masonry walls cannot be disassembled and reused. The installations of doors, windows, and anti-burglary metal components are installed in such a way that some demolition of parts of the building must be carried out before the components can be removed. Such partial demolitions can impair the components and render them unusable subsequently. This also applies to building services installations water, drainage, and air-conditioning services. This is particularly challenging during routine maintenance work especially for building services.

In terms of construction methods adopted, the *in situ* construction methods characterized most of the building construction activities executed under the program. Hence, labor-intensive methods rather than technology-intensive methods were deployed in the construction of the buildings. Even though labor-intensive methods tend to be advantageous in terms of creating employment opportunities

Figure 4.
Hoisting of Prefabricated Slabs. Source: [91].

for the population, it often leads to so much waste that runs against the tenets of circular economy.

6.2 Use of renewable, bio-based, and waste-based materials

Specification of materials for the project favored conventional building materials as indicated in **Table 3**. Even though lightweight composite materials are lighter and faster to erect as partition walls, none were used for the buildings. High reliance on cement as a major building material makes the buildings less energy-efficient in terms of the embodied energy content of cement and cement-based materials. This can be mitigated by reducing the quantity of cement used in housing development by using cement substitutes usually referred to as supplementary cementitious materials (SCMs). There is also the need to adopt construction methods that would minimize the use of Portland cement and other energy-intensive materials. A number of alternative building materials that combine low embodied energy with speed of erection have been identified in Nigeria and they include interlocking bricks for mortar-less wall construction, expanded polystyrene panels for internal walls, and composite building panels for walls and ceilings. Similarly, bio-based materials most of which are by-products of agricultural processing have been found to be very useful, though poorly deployed in the Nigerian context. Apart from timber, no bio-based material was specified in the buildings.

6.3 Adoption of renewable energy

There is no planned integration of renewable energy whether at micro- or mini-grid levels. However, energy-efficient appliances were installed in the apartments. Given the household energy situation characterized by low access and increasing cost of available ones, the residents have on their own started the introduction of renewable retrofits in the buildings. However, this has not proceeded on an organized basis to accommodate all occupants. Going forward, renewable energy integration into the planning and implementation of the housing projects should be considered a priority so the buildings do not go into early obsolescence. The deployment of passive design principles is not profound as observed previously.

In order to make renewable energy integration effective, passive strategies must be deployed fully. Building orientation, cross-ventilation, and use of shading devices and elements among others should be brought to the front burner.

6.4 Deployment of passive and regenerative strategies

Passive design strategies such as building orientation, cross-ventilation, vegetative cover, shading devices, and roof overhangs can be better deployed in the buildings. The atrium can be made to play a bigger environmental role. The design objective was aimed at maximizing the use of the site due to the development pressure on land. Green spaces were few and far between, while the buildings are closely spaced together to increase density. Lagos has a small build-able land area relative to the population and housing needs. As a result, densification is a deliberate development policy of the government. Such dense developments would have benefited from the installation of green roofs and other similar green infrastructure installations. However, no green roofs or green facades are incorporated. Hence, the ecosystem regeneration strategies were not substantially deployed. The roof adopted a lean-to design, which helped to reduce the roof footprint, thereby conserving usage of roofing materials.

7. Conclusions and recommendation

This chapter underscored the importance of circular economy as the preferred direction for all human activities. The advantage of circular economy with particular reference to the built environment was highlighted. Specifically, given the impact of the built environment in terms of resource use and waste generation, circular economy principles were adjudged to be the panacea. Given the importance of housing in the overall built environment, circular economy offers positive prospects. The relevance of circular strategies in housing delivery within a developing country context such as Nigeria becomes more apparent. With housing deficit running into millions, the resource implication of bridging the deficit is huge. Adoption of circular design strategies can effectively mitigate the negative resource implication of bridging the housing deficit.

The study also evaluated an urban public housing program in Lagos, Nigeria's most populous and urbanized city. It was found that opportunities exist for the adoption of circularity in housing delivery given the huge housing need. Adoption of modern methods of construction, which will incorporate the circular design principles of standardization, prefabrication, modularity, reuse, and design for disassembly, was identified as good approach to reducing the housing deficit. However, the prevailing building procurement process followed a rudimentary process of wet or *in situ* construction. Similarly, the design of the buildings did not incorporate innovative processes that would facilitate modern methods of construction. In order to move the housing delivery process toward circularity in the study area, modern construction technology should be adopted. As an active player in the housing delivery process, the public sector can play an important role in the uptake of modern technology in building design and construction process. In this respect, the building procurement process should benefit from digital technology.

The importance of renewable materials was also highlighted in the chapter. Renewable materials can be brought into the material mix due to the high cost of housing procurement using conventional materials. A number of renewable materials are found in the study area that can be used for housing development. There are materials such as SCMs that can reduce the quantity of Portland cement deployed in the construction of the buildings. Incidentally, these SCMs can be obtained from byproducts of both industrial processes and as wastes from agricultural processing. In this respect, wastes from other processes are utilized as inputs into building materials for housing development, thereby promoting circularity. Also, the use of mortar-less interlocking blocks can reduce the use of high-impact Portland cement. Interlocking clay bricks stabilized with cement have also been found to be very useful in the study context.

Energy use accounts for a large proportion of resources used in housing, especially at the use/operational level. Renewable energy use is on the increase, driven by the need for cleaner and low-impact energy. As a result, public housing projects are increasingly adopted renewable energy to mitigate the waste associated with conventional energy supply. Even though renewable energy was not factored into the housing projects, residents and building occupiers have commenced energy retrofits using solar photovoltaic installations.

Hence, going forward, energy efficiency strategies that incorporate solar photovoltaic installations with other building components are desirable. Mini-grid PV networks can be considered for estates.

Finally, the adoption of passive and regenerative principles in housing design should be emphasized. Housing design should conform to context in order to maximize natural attributes through passive design principles. This can help to reduce overall environmental impact of the buildings and reduce the need for the use of high-impact equipment and accessories. In a similar vein, green infrastructure is a way of promoting regenerative principles in housing design. The preponderance of hard landscaping in the housing programs informs the need for green infrastructure incorporation. However, given the limited land area available in the study area for housing development, conventional green infrastructure may not be feasible. Hence, while increasing occupancy per unit area in response to estimated deficit, it is desirable to explore the adoption of green roofs as a way of recapturing the natural green areas displaced by the construction of buildings and other infrastructure.

Acknowledgements

The authors acknowledge the support provided by Covenant University Centre for Research, Innovation and Discovery CUCRID toward the research leading to this publication.

Conflict of interest

The authors declare no conflict of interest.

Author details

Isidore C. Ezema*, Taofeek A. Suleman and Regina K. Okorigba
Department of Architecture, Covenant University, Ota, Nigeria

*Address all correspondence to: isidore.ezema@covenantuniversity.edu.ng

References

[1] Kedir F, Hall DM. Resource efficiency in industrialized housing construction – A systematic review of current performance and future opportunities. Journal of Cleaner Production. 2021;**286**:125443. DOI: 10.1016/j.jclepro.2020.125443

[2] Marchesi M, Tweed C, Gerber D. Applying circular economy principles to urban housing. IOP Conference Series: Earth and Environmental Sciences. 2020;**588**:052065. DOI: 10.1088/1755-1315/588/5/052065

[3] Silva MF. Another way of living: The prefabrication and modularity toward circularity in the architecture. IOP Conference Series, Earth and Environmental Sciences. 2020;**588**(4):042048. DOI: 10.1088/1755-1315/588/4/042048

[4] Ezema IC, Okorigba RK. Life circle energy and Co2 analysis for environmental sustainability of the Nigerian housing stock. Architecture Research. 2022;**12**(2):27-37

[5] WEF - World Economic Forum, Ellen MacArthur Foundation, McKinsey and Company. Towards the Circular Economy: Accelerating the Scale-up Across Global Supply Chains. Switzerland: World Economic Forum; 2014

[6] Xue K, Hossain MU, Liu M, Ma M, Zhang Y, Hu M, et al. BIM integrated LCA for promoting circular economy towards sustainable construction: An analytical review. Sustaianbility. 2021;**13**(3):1-21. DOI: 10.3390/su13031310

[7] Akhimien NG, Latif E, Hou SS. Application of circular economy principles in buildings: A systematic review. Journal of Building Engineering. 2021;**38**(102041):1-42. DOI: 10.1016j. jobe.2020.102041

[8] Fatourou-Sipsi A, Symeonidou I. Designing [for] the future: Managing architectural parts through the principles of circular economy. IOP Conference Series Earth and Environmental Sciences. 2021;**899**(012014):1-8. DOI: 10.1088/1755-1315/899/1/012014

[9] EMF - Ellen MacArthur Foundation. Growth within: A circular economy vision for a competitive Europe. 2015; Available from: https://ellenmacarthurfoundation.org/growth-within-a-circular-economy-vision-for-a-competitive-europe

[10] Akanbi LA, Oyedele LO, Omoteso K, Bilal M, Akinade OO, Ajayi AO, et al. Disassembly and deconstruction analytics system (D-DAS) for construction in a circular economy. Journal of Cleaner Production. 2019;**223**:386-396. DOI:10.1016/j.jclepro.2019.03.172

[11] Attia S, Santos MC, Al-Obaidy M, Baskar M. Leadership of EU member states in building footprint regulations and their role in promoting circular building design. IOP Conference Series. Earth and Environmental Science. 2021;**855**(012023):1-13. DOI: 10.1088/1755.1315/855/1/012023

[12] Ekpo A. Housing deficit in Nigeria: Issues, challenges and prospects. Central Bank of Nigeria Economic and Financial Reviews. 2019;**57**(4):201-2220

[13] Aremu AS, Olukanni DO, Mokuolu OA, Lasode AO, Ahove MA, Ojowuro OM. Circular Economy: Nigeria Perspective. Springer Nature; 2020. pp. 279-297

[14] Olukanni DO, Nwafor CO. Public-Private sector involvement in providing efficient solid waste management

services in Nigeria. Recycling. 2019;**4**:19. DOI: 10.3390/recycling4020019

[15] Olukanni DO, Pius-Imue FB, Joseph SO. Public perception of solid waste management practices in Nigeria. Recycling. 2020;**5**:8. DOI: 10.3390/recycling5020008

[16] Omole DO, Jim-George T, Akpan VE. Economic analysis of wastewater reuse in Covenant University, IOP Conf. Series: Journal of Physics Conf. Series. 2019:1299. DOI: 10.1088/1742-6596/1299/1/012125

[17] Ellen MacArthur Foundation [EMF]. What is a Circular Economy? Isle of Wight: Ellen MacArthur Foundation; 2017. Available from: https://ellenmacarthurfoundation.org/topics/circular-economy-introduction/overview

[18] Etkins P, Domenech T, Drummond P, Bleischwitz R, Hughes N, Lotti L. The circular economy: What, Why, How and Where, Background paper for an OECD/EC Workshop on 5 July 2019 within the workshop series. In: Managing Environmental and Energy Transitions for Regions and Cities. Paris: OECD

[19] Weavabel UK. The role of circular design in reducing environmental impact. 2022. Available from: https://www.weavabel.com/blog/the-role-of-circular-design-in-reducing-environmental-impact

[20] Radjou N, Prabhu J. Frugal Innovation: How to do More with Less. London: Profile Books Ltd; 2015

[21] Edwards B. Rough Guide to Sustainability. 1st ed. London: RIBA Publications; 2002

[22] Ness DA, Xing K. Towards a resource-efficient built environment: A literature review and conceptual model. Journal of Industrial Ecology. 2017;**21**:572-592

[23] UNEP - United Nations Environment Programme. Global Status Report for Buildings and Construction: Towards a Zero-emission, Efficient and Resilient Buildings and Construction Sector. Nairobi: UNEP; 2020. p. 2020

[24] Forsberg A, Von-Malmborg F. Tools for environmental assessment of the built environment. Building and Environment. 2004;**39**:223-228

[25] Rahla KM, Braganca L, Mateus R. Obstacles and barriers for measuring building's circularity. IOP Conference Series: Earth and Environmental Sciences. 2019;**225**(012058):1-8. DOI: 10.1088/1755-1315/225/1/012058

[26] Attia S, Al-Obaidy M. Design Criteria for Circular Buildings. In: IBA Crossing Boundaries Conference. Parkstad, The Netherlands: IBA; 24-25 March 2021. Available from: https://www.researchgate.net/publication/350664011_Design_criteria_for_circular_buildings

[27] EMF - Ellen MacAuthor Foundation. Project overview: An approach to measuring circularity. 2015; Ellen MacArthur Foundation. DOI: 10.1016/j.giq.2006.04.004

[28] Brandstrom J, Saidani M. Comparison between circularity metrics and LCA: A case study of circular economy strategies. Journal of Cleaner Production. 2022;**371**:133537. DOI: 10.1016/j.jclepro.2022.133537

[29] Drager P, Letmathe P, Reinhart L, Robineck F. Measuring circularity: Evaluation of the circularity of construction products using the OKOBAUDAT database. Environmental Science Europe. 2022;**34**(13):1-16. DOI: 10.1186/s12302-022-00589-0

[30] Platform CB'23. Measuring circularity: Working agreements for

circular construction, Version 2. 2020; Available from: https://platformcb23.nl/images/downloads/Platform_CB23_Guide_Measuring_circularity_2.0.pdf

[31] Rigamonti L, Mancini E. Life cycle assessment and circularity indicators. International Journal of Life Cycle Assessment. 2021;**26**:1937-1942

[32] WBCSD - World Business Council for Sustainable Development. Circular transition indicators, V1.0: Metrics for business, by business. 2020. Available from: https://www.wbcsd.org/Programs/Circular-Economy/Metrics-Measurement/Resources/Circular-Transition-Indicators-V1.0-Metrics-for-business-by-business

[33] KPMG. Circular economy: A framework to get you started. 2020. Available from: https://www.kpmg.com/sustainability

[34] Saade M, Erradhouani B, Pawlak S, Appendino F, Peuportier B, Roux C. Combining circular and LCA indicators for the early design of urban projects. International Journal of Life Cycle Assessment. 2022;**27**(1):1-19. DOI: 10.1007/s11367-021-02007-8

[35] Weidemann SG, Nguyen Q, Clarke SJ. Using LCA and circularity Indicators to measure the sustainability of textiles: Examples of renewable and non-renewable fibres. Sustainability. 2022;**14**(16683):1-14

[36] Mannan M, Al-Ghamdi SG. Complementing circular economy with life cycle assessment: Deeper understanding of economic, social and environmental sustainability. In: Circular Economy and Sustainability. Vol. 1. Amsterdam: Elsevier Inc.; 2022. DOI: 10.1016/B978-0-12-819817-9.00032-6

[37] Van Stijn A, Eberhardt LCM, Jansen BW, Meijer A. A circular economy life cycle assessment (CE-LCA) model for building components. Resources, Conservation & Recycling. 2021;**174**(105683):1-34. DOI: 10.1016/j.resconrec.2021.105683

[38] McGraw Hill Construction. Prefabrication and Modularization: Increasing Productivity in the Construction Industry, Smart Market Report. New York: McGraw Hill; 2011

[39] Page I, Norman D. Prefabrication and Standardisation Potential in Buildings, Study Report SR312, Branz, New Zealand, 2014

[40] Eberhardt LCM, Birkved M, Birgisdottir H. Building design and construction strategies for a circular economy. Architectural Engineering and Design Management. 2022;**18**(2):93-113. DOI: 10.1080/17452007.2020.1781588

[41] Cheng Z, Tang S, Liu H, Lei Z. Digital technologies in offsite and prefabricated construction: Theories and applications. Buildings. 2023;**13**:63. DOI: 10.3390/buildings13010163

[42] Trzcińska M. Add, transform, and utilize. Possibilities of applying druot, lacaton, and vassal's modernization strategies and solutions in Polish large-panel housing estates. Land. 2021;**10**(12):1-20. DOI: 10.3390/land10121308

[43] Oloto EN, Iweka ACO. Private developers perceived usefulness of the prefabrication technology in Nigeria's housing development industry. Kufa Journal of Engineering. 2022;**13**(4):25-38

[44] Adeyemo AF, Kolawole JT. Effect of modularisation on cost and time performance of building products in Lagos state. Nigeria, Journal of Environmental Design and Management. 2016;**8**(1&2):155-162

[45] Çetin S, Gruis V, Straub A. Towards circular social housing: An exploration of practices, barriers, and enablers. Sustainability. 2021;**13**(4):1-24. DOI: 10.3390/su13042100

[46] Durmisevic E. Reversible Building Design Guidelines and Protocols, Working Paper WP3/10/UT. Netherlands: University of Twente; 2018. Available from: https://www.bamb2020.eu/wpcontent/uploads/2018/12/Reversible-Building-Design-guidelines-and-protocol.pdf

[47] Kröhnert H, Itten R, Stucki M. Comparing flexible and conventional monolithic building design: Life cycle environmental impact and potential for material circulation. Building and Environment. 2022;**222**:1-13. DOI: 10.1016/j.buildenv.2022.109409

[48] Cambier C, Galle W, De Temmerman N. Expandable houses: An explorative life cycle cost analysis. Sustainability. 2021;**13**(12):1-28. DOI: 10.3390/su13126974

[49] Askar R, Bragança L, Gervásio H. Adaptability of buildings: A critical review on the concept evolution. Applied Sciences. 2021;**11**(10):1-32. DOI: 10.3390/app11104483

[50] Ipsen KL, Pizzol M, Birkved M, Amor B. How lack of knowledge and tools hinders the eco-design of buildings—A systematic review. Urban Science. 2021;**5**(1):20. DOI: 10.3390/urbansci5010020

[51] Koster M, Schrotenboer I, Van der Burgh F, Dams B, Jacobs L, Versele A, et al. White paper: Five essentials for successful circular bio-based construction initiatives. Circular Bio-based Construction Industry (CBCI). 2020

[52] Kayacetin NC, Versele A. A circular and bio-based renovation strategy for low-income neighbourhoods. IOP Conference Series: Earth and Environmental Sciences. 2022;**1078**:012080. DOI: 10.1088/1755-1315/1078/1/012080

[53] Olorunnisola AO. Development of sustainable building materials from agro-industrial wastes in Nigeria. In: Sustainable Construction and Building Materials. London: IntechOpen; 2019. DOI: 10.5772/intechopen.81662

[54] Ayeni DA, Omeiza VO. Investigating design strategies for regenerative development. European Journal of Engineering and Technology Research. 2021;**6**(1):87-94. DOI: 10.24018/ejers.2021.6.1.1987

[55] Brown M, Haselsteiner E, Apro D, Kopeva D, Luca E, Pulkkinen K, et al. Sustainability, Restorative to Regenerative. COST Action CA16114RESTORE, Working Group One Report: Restorative Sustainability. Brusels: COST; 2018

[56] Morseletto P. Restorative and regenerative: Exploring the concepts in the circular economy. Journal of Industrial Ecology. 2020;**24**(4):763-773. DOI: 10.1111/jiec.12987

[57] Petrovski AA, Pauwels E, González AG. Implementing regenerative design principles: A refurbishment case study of the first regenerative building in Spain. Sustainability. 2021;**13**(4):1-18. DOI: 10.3390/su13042411

[58] Akhimien NG, Laif E. Incorporating circular economy into passive design strategies in tropical Nigeria. International Journal of Economics and Management Engineering. 2019;**13**(10):1380-1385

[59] Khan AM, Osinska M. Energy consumption under circular economy conditions in the EU countries. Energies. 2022;**15**:7839. DOI: 10.3390/en15217839

[60] Lembi JJ, Akande OK, Ahmed S, Emechebe LC. The drivers for low energy materials application for sustainable public housing delivery in Nigeria. Landscape Architecture and Regional Planning. 2021;**6**(2):19-24. DOI: 10.11648/j.larp.20210602.11

[61] Ugochukwu S, Okongwu I. Appraising the problems militating against effective housing budgeting practice in Akwa Ibom State. International Journal of Scientific and Research Publications. 2021;**11**(8):571-582. DOI: 10.29322/IJSRP.11.08.2021.p11670

[62] Olubi AR, Aseyan BS. Sustainable housing delivery for the urban poor in Nigeria. International Journal of Civil Engineering, Construction and Estate Management. 2021;**10**(1):21-34

[63] Yakubu MJ, Muhammad MS, Adebenga A, Gambo MJ. Effective civil servant's housing ownership in Bauchi metropolis. Journal of Environmental Design and Construction Management. 2021;**21**(4):35-54

[64] Olubi AR, Fadamiro JA. Implications of urbanisation on sustainable cities development in Nigeria. European Journal of Development Studies. 2022;**2**(4):4-13. DOI: 10.24018/ejdevelop.2022.2.4.94

[65] Ogunbayo B, Aigbavboa C, Thwala W. Institutional evaluation of public and private partnerships relevant contributions in housing delivery system, human factors in architecture. Sustainable Urban Planning and Infrastructure. 2022;**58**:370-378. DOI: 10.54941/ahfe1002374

[66] Moore EA. Addressing housing deficit in Nigeria: Issues, challenges and prospects. CBN Economic and Financial Review. 2019;**57**(4):201-222

[67] Okoye NBCD, Onyegiri I, Okafor M. Effect of architectural design characteristics affecting design simplicity on affordability improvement of core housing schemes. Journal of Advanced Research and Multidisciplinary Studies. 2021;**1**(1):26-59

[68] Olanrewaju SBO, Okedare DKO, Fagbohun JO. Prospect of low-cost housing delivery for middle income earners in Nigeria. The International Journal of Engineering and Sciences. 2021;**10**(7):48-54

[69] Othman MS. Citizens' perception of public housing delivery in Gombe metropolis. Path of Science. 2021;**7**(10):2001-2006. DOI: 10.22178/pos.75-4

[70] Alabi B, Fapohunda J. Effects of increase in the cost of building materials on the delivery of affordable housing in South Africa. Sustaianbility. 2021;**13**(1772). DOI: 10.3390/su13041772

[71] Akande O, Akoh S, Francis B, Odekina S, Eyigege E, Abdulsalam M. Assessing the potentials of low impact materials for low energy housing provision in Nigeria. Journal of Sustainable Construction Materials and Technologies. 2021;**6**(4):156-167. DOI: 10.14744/jscmt.2021.04

[72] Ghisellini P, Ripa M, Ulgiati S. Exploring environmental and economic costs and benefits of a circular economy approach to the construction and demolition sector. A literature review. Journal of Cleaner Production. 2018;**178**:618-643. DOI: 10.1016/j.jclepro.2017.11.207

[73] Torgautov B, Zhanabayev A, Tleuken A, Turkyilmaz A, Mustafa M, Karaca F. Circular economy: Challenges and opportunities in the construction sector of Kazakhstan. Buildings. 2021;**11**(11):1-19. DOI: 10.3390/buildings11110501

[74] Ezennia IS. Insights of housing providers' on the critical barriers to sustainable affordable housing uptake in Nigeria. World Development Sustainability. 2022;**1**:100023. DOI: 10.1016/j.wds.2022.100023

[75] BuyLetLive. Three numbers that show you how bad the housing deficit in Lagos really is. 2022. Available from: https://nairametrics.com/2022/02/26/three-numbers-that-show-you-how-bad-the-housing-deficit-in-lagos-really-is/

[76] Patunola-Ajayi OL. An appraisal of Lagos state housing delivery strategies: A remedy to Lagos state housing challenges. African Scholar Journal of Environmental Design & Construction Management. 2020;**19**(4):1-22

[77] Jeje B. Green Buildings and Low Energy Houses as Adaptation Tools for Climate Change in Lagos State, Lead paper, 5th Lagos State Climate Change Summit, 13-15 March, 2013. Nigeria: Lagos; 2013

[78] Alufohai AJ. The Lagos State 2010 mortgage law and the supply of housing. In: Paper Presented at the International Federation of Surveyors FIG Working Week - Environment for Sustainability. Abuja, Nigeria: FIG; 6-10 May 2013

[79] Lagos State Ministry of Housing. Rent - to - Own: All you need to know. 2017. Available from: https://lagosstate.gov.ng/wpcontent/uploads/2017/07/BOOKLET-HOUSING.pdf

[80] Aboginije A, Aigbavboa C, Thwala W. A holistic assessment of construction and demolition waste management in the Nigerian construction projects. Sustainability. 2021;**13**:6241. DOI: 10.3390/su13116241

[81] Kolo SJ, Rohimian FP, Goulding JS. Offsite manufacturing construction: A big opportunity for housing delivery in Nigeria. Procedia Engineering. 2014;**85**(2014):319-327

[82] Dixon-Ogbechi BN, Adebayo KA. Application of AHP model to determine prefab housing adoption factorsfor developers in Lagos. International Journal of the Analytic Hierarchy Process. 2020;**12**(2):297-327. DOI: 10.13033/ijahp.v12i2.635

[83] Marut JJ, Alaezi JO, Obeka IC. A review of alternative building materials for sustainable construction towards sustainable development. Journal of Modern Materials. 2020, 2020;**7**(1):68-78. DOI: 10.21467/jmm.7.1.68-78

[84] Olorunnisola AO. Development of sustainable building materials from agro-industrial wastes in Nigeria. 2019. DOI: 10.5772/intechopen.81662

[85] Fapohunda C, Kilani A, Adigo B, Ajayi L, Famodimu B, Oladipupo O, et al. A review of some agricultural wastes in Nigeria for sustainability in the production of structural concrete. Nigerian Journal of Technological Development. 2021;**18**(2):1-12. DOI: 10.4314/njtd.v18i2.1

[86] Elinwa UK, Ogbeba JE, Agboola OP. Cleaner energy in Nigeria residential housing. Results in Engineering. 2021;**9**:100103. DOI: 10.1016/j.rineng.2020.100103

[87] Abbakyari M, Taki A. Passive Design Strategies for Energy Efficient Housing in Nigeria. PLEA 2017. Proceedings, Passive and Low Energy Architecture Conference, Edinburgh. 2017

[88] Ochedi ET, Taki A. Energy efficient building design in Nigeria: An assessment of the effects of the sun on energy consumption in residential buildings. Journal of Engineering and Architecture. 2019;**7**(1):1-18. DOI: 10.15640/jea.v7n1a1

[89] Ochedi ET, Taki A. A framework approach to the design of energy efficient residential buildings in Nigeria. Energy and Built Environment. 2022;**3**(2022):384-397

[90] Ogunleye OS, Coenen F, Hoppe T. Stakeholder perspectives on community energy contributing to the use of renewable energy sources and improving energy security in Nigeria. Energies. 2022;**15**:7390. DOI: 10.3390/en15197390

[91] Ezema IC. Innovation for sustainability in the design and construction of public housing: The case of Lagos home ownership and mortgage scheme. In: Laryea S, Ibem E, editors. Proceedings 8th Construction Industry Development Board (cidb) Postgraduate Conference, 10-11 February 2014. South Africa: University of the Witwatersrand, Johannesburg; 2014. pp. 357-370

[92] Google Earth. Available from: https://earth.google.com/web/search/Igando,+Lagos/@6.5346654,3.2444,19.958148a,18476.78553373d,35y,0h,0t,0r/data=CngaThJICiUweDEwM2I4ZmYyMmViYzk4OGQ6MHg0ZjU5ODE2MTM5OGIyNjAwGQMlBRbAJBpAIcZBU8CH_AlAKg1JZ2FuZG8sIExhZ29zGAIgASImCiQJwcVdF6TnGUARtBcnZ2jWGUAZuRxCr63oCkAhMw-CcMXMCkA

Section 2

Addressing the Humanitarian Challenges of Climate Change

Chapter 3

Experimental Living and Housing Forms: Cities of the Future as Sustainable and Integrated Places of Food Production

Alessandra Battisti, Alberto Calenzo and Livia Calcagni

Abstract

Population growth and urbanization are progressively leading to an increase in the global food demand within cities resulting in a rise in global greenhouse gas (GHG) emissions, land consumption, resource depletion, and social tensions. The key challenge for future decades is to feed a growing population in an ethical and socially, economically, environmentally sustainable way. Traditional city and housing models are no longer capable of providing a compelling solution. The urgency of providing dynamic responses in terms of integrated urban solutions must coexist with a medium- to long-term perspective in which production is gradually embedded within the urban structure. Since the relationship between places of production and consumption is a critical node in food policy, it is essential to strengthen this link within a more globalized and interconnected economy. This essay investigates two different strategies: on the one hand, agri-cities and communities as an experimental social-business model that places the production once again at the center of housing design, and on the other floating potential for food production in delta and coastal cities, as a zero-land footprint strategy. In both approaches, cities and the way they work must be reimagined with a view to making them locally productive and globally connected.

Keywords: productive city, agri-city, land scarcity, urban farming, floating potential

1. Introduction

Changing global housing conditions due to recent climate hazards, migration, war, pandemic events, and globalization are further exacerbated by the rapidly unfolding economic global crises [1]. Today, we are witnessing two seemingly opposing phenomena: on the one hand, an inevitable increase in the world's population and on the other, a constant inexorable aging and decrease in the Western population [2]. Global population growth implies an increase in the demand for food production, especially that related to animal-derived protein consumption, which will have a significant impact on the whole production chain. According to the World Resources Institute, the global consumption of meat and dairy products is expected to rise by about 70%

between 2010 and 2050, with beef consumption increasing by more than 80% [3]. The United Nations (UN) Food and Agriculture Organization (FAO) projects that annual global meat production will rise from 228 to 463 million tons by 2050 to meet the rising demand, with bovine production expected to increase from 1.5 to 2.6 million tons, and sheep and goat products from 1.7 to 2.7 million tons [4]. The global food system, besides being one of the greatest threats to the planet's biodiversity, accounting for 80% of global species and habitat loss, is also one of the most polluting sectors [5]. Agriculture accounts for 24% of worldwide greenhouse gas (GHG) emissions [6]. Intensive livestock production is a major source of GHG emissions in this sector, accounting for 14.5% of total emissions [7]. Furthermore, according to the UN report "The World Population Prospects 2019: Highlights," the global population will reach 9.7 billion people by 2050, rising to nearly 11 billion by the end of the century [8]. Without a radical change, these emissions are expected to rise as the world's population and food demand increase. Traditional city and housing models are no longer capable of providing a compelling response to these emerging and unavoidable social and environmental challenges [9]. The threats posed by climate and socioeconomic changes, as well as the resulting alterations in environmental balances, require immediate mitigation and adaptation measures. In urban areas, these strategies affect not only land use, but also lifestyles and production. Half of the world's population and three-quarters of Europe's population live in urban areas. Cities are key contributors to rising environmental pressures, with significant withdrawals of natural resources and pollution emissions. Moreover, increasing environmental pressure has serious implications for public health and safety. As a result, cities play a crucial role in achieving the sustainable development goals (SDGs) [10]. Goal 11 of the UN 2030 Agenda [11] lays the groundwork for integrated sustainable urban solutions. The SDG 11 declaration statement emphasizes the most compelling need for an integrated and systematic management of all potential dimensions of contemporary life on earth [12]. According to the UN, the world is falling behind in reaching the SDGs, and the efforts made thus far are completely inadequate. A radical shift in lifestyles is needed to achieve these goals in the shortest amount of time. The urgency of providing dynamic responses [13] that are adaptable to unpredictable challenges [14] has largely contributed to the development of renewed housing cultures that are more inclusive and shared [15]. The dynamic responses and new housing cultures must coexist with a medium- to long-term view, in which the development of creative innovative community forms serves as a means of empowering end-users. This radical is supported by citizens' increasing awareness of the use and importance of autonomous and decentralized food and energy production models.

1.1 The challenge of combining housing and food production

Food and related activities—production, processing, distribution, consumption, and postconsumption—are key contributors to urban-scale unsustainability in environmental, social, and economic terms [16]. The current food production system has optimized food supply chain management, lowering product costs but with major negative consequences [17]. These impacts not only affect the natural environment (loss of agricultural and natural biodiversity; increased competition for land, land grabbing, and new forms of food colonialism), but especially cities, which are becoming increasingly distant—not only physically—from places of production. The main related issues include environmental pollution (waste, land consumption; reliance on fossil fuels and GHG emissions, traffic, and water consumption for production), social tensions, food crises, and the soaring increase in diseases related to obesity and unhealthy eating habits, particularly among low-income groups [18].

Cities are currently considered as the major engines of economic progress. As a result, rural places are assigned a marginal role. This polarity leads to the widening of the gap between cities and their territorial contexts (including supply), disrupting the material and immaterial flows that connect them to rural areas [19]. Already in the nineteenth century, Marx had theorized a fracture in the metabolic interaction between humanity and the rest of nature as a result of capitalist agricultural production and the rising divide between town and country [20]. This fracture is known as the metabolic rift. More precisely, this rift results in the loss of biodiversity, depletion of natural resources, and environmental degradation in urban environments [21], drawing attention to the need to rethink sustainable local agri-food systems and thus redefine relations and a balance between the city and the countryside.

Since the relationship between places of production and consumption, between city and rural, metropolitan and peri-urban areas, is a critical node in food policy [22], it is essential to strengthen this link within a more globalized and interconnected economy. The significance of physical and organizational proximity in different social, cultural, and economic relationships must be rediscovered. The emphasis on production, the traditional urban–rural dualism, and an increasingly global and de-territorialized agri-industrial system has resulted in the disappearance of food from reflections on urban development, after having shaped and molded the form and substance of cities for centuries [23]. Indeed, in cities, there is (little) awareness of the act of consuming, whereas the other phases of the supply chain tend to be overlooked [24].

To reverse this trend, as early as 1997 the European Commission began advocating for a "more versatile, sustainable, competitive, and widespread European model of agriculture." The Farm to Fork EU Strategy is at the heart of the European Green Deal and aims to accelerate our transition to a sustainable food system. It addresses comprehensively the challenges of sustainable food systems and recognizes the inextricable links between healthy people, healthy societies, and a healthy planet. This model has found application not only in a wide variety of farm facilities, land cultivation systems, and range of products, but also in the spread of new multifunctional rural and urban settlements. The new patterns of productive settlement have led us to rethink the role of agriculture in urban areas and to refine the design and implementation of experimental housing-productive settlements, characterized by a long-term gradual transformation of living and housing models.

Food Trails is a European Union (EU)-funded Horizon 2020 project, bringing together a consortium of 19 European partners, including 11 cities, 3 universities, and 5 organizations. The project is rooted in the Milan Urban Food Policy Pact (MUFPP), an international mayors' agreement, and aims to enable cities to reimagine, develop, and implement sustainable, healthy, and inclusive food policies. Each partner city runs a pilot project, a "Living Lab," which seeks to codesign and co-implement food actions integrated with other local sectoral works and aligned with the Farm to Fork EU Strategy and the priorities of the EU-FOOD 2030 Policy: nutrition, climate, circularity, and innovation. Grenoble-Alpes Metropole, for instance, has a unique food system linked to its geography. Around 15% of its mountainous territory is covered by farmable land, and the local government sought to capitalize on it to improve diets, sustainability and create short food chains. The Metropolitan Agricultural Strategy 2015–2020 implemented by the Metropole aims to re-territorialize its food system, promoting sustainable and high-quality farming in rural municipalities while connecting them to other metropole cities via short supply chains, supporting farmers in adapting agriculture and food production to climate change, and reducing the environmental impact of local horticulture in order to reduce GHG emissions by 2050. Moreover, it seeks to develop a participatory scheme from a Food and Agriculture Policy and Strategy (FAPS) toward a Common Food Policy.

Considering the current transformations cities are experiencing, the key challenge for future decades is to feed a growing population in an ethical and socially, economically, environmentally sustainable way [25]. Therefore, the search for an alternative food paradigm through food policies based on relocation, critical consumption (fresh, local, organic), and nutrition education is necessary. Relocation does not mean achieving complete food self-sufficiency, but rather producing locally a greater portion of the basic food demand. This purpose underlies policies such as urban and peri-urban farm protection and promotion, alternative food networks, optimization of distribution and logistics stages in a short supply chain perspective, and public procurement.

The EU Joint Research Centre report "Farmers of the Future" [26] reflects on the future of agriculture in the coming decades and what characteristics farmers will have in 2040. It highlights "the emergence of more diverse and experimental agriculture models to address environmental challenges and respond to different consumption patterns."

This change toward a wider range of housing types integrated with extensive agricultural facilities, in addition to having strong implications for governance, requires adaptation of farming and livestock systems to local specificity. Three factors must be considered in terms of producing food for consumption: assessment of socio-ecological changes, interpreted as typological-spatial variation in housing and settlements; recognition of multifunctionality and supply of public goods as intrinsic tasks associated to housing; and creation of a suitable governance framework capable of systematizing all aspects of production (energy, labor, agriculture, and livestock) integrated into experimental housing models.

2. Food-producing communities as dynamic urban laboratories for sustainable living

Modern agroecology proposes a new multidisciplinary, intersectoral, and multiscalar approach to redefining the relationship between production, cities, and land, both by providing a more current vision of agriculture, which influences the development of new management, monitoring, and planning tools, and by offering a different perspective to restructure the relationship between agriculture and society [27]. In contrast to the assumptions of a standard agricultural management and the concept of a one-size-fits-all agricultural model, experimental agri-urban models developed in recent decades emphasize the importance of the management of all resources involved in agricultural production processes and emphasize the need to promote diffuse management and production systems. Such models ascribe to food production not only the ability to sustainably cope with the growing food demand from a short supply chain perspective—thus reducing critical factors associated with long-distance transportation such as GHG emissions and food waste—but also the ability to improve living conditions by increasing the degree of multifunctionality, supply of public goods, ecosystem quality, and even microclimate conditions. Indeed, bringing nature into cities through urban farming not only provides high-quality zero-mile products and a variety of social functions, but it also has the potential to mitigate many of the polluting phenomena that affect highly urbanized territories. Food-producing cities are livable cities: they create new connections on a social, ecological, and economic level.

Against this backdrop, cities must seize the opportunity to renew themselves through the adoption of a new business model that places the production once again at the center of housing design. Not just production of fresh food, materials, and energy, but a production of intangible value in the transition to a more sustainable future. As a result, cities and the way they work must be reimagined with a view to making them locally productive and globally connected.

In recent decades, the use of the term agritecture has grown in popularity as a means of transforming and reinventing the food supply of future cities. In the productive city concept, places of agriculture or food production can shape new peri-urban contexts or find space along the edges of sub-urbanization or within established urban fabrics, in existing buildings, in public spaces, in residual spaces, even on terraces and courtyards, in a comprehensive redesign of the metropolitan landscape. Several experts highlight how awareness on the interaction between urban agriculture and contemporary urban space has increased in recent years [28, 29]. Areas intended for farming are being reclaimed and regenerated in abandoned or in-transition urban and peri-urban contexts. Agritecture takes several forms, especially as innovative agricultural models, integrated into buildings. Applications are mainly classified as follows: hanging gardens and/or intensive green roof systems designed to grow fruits and vegetables using soil-based production methods; rooftop greenhouses, "vertical farms," "plant factories," or "indoor farms" that use multistory vertical systems for food production that rely on controlled environment agriculture (CEA) methods that aim to optimize crop growth and space occupied through above-ground growing techniques such as hydroponics, aquaponics, and aeroponics.

Thus, food production takes place in close proximity to consumers, with the possibility of using urban waste as an input for food production in a circular system. The recovery of rainwater, wastewater, waste heat, and organic waste, for example, provides a valuable opportunity to supply water, energy, and nutrients to food production systems while reducing the load on the respective urban drainage and treatment systems.

On the other hand, the progressive recognition of the importance of food in urban development patterns, as well as increased awareness of the impact and externalities of the agri-food system, particularly at the socioeconomic level, have led local governments in recent years to regain responsibility for food and actively engage in the development of urban food policies. Food policies, in fact, place (or relocate) food at the center of urban policy agendas, capitalizing on existing experiences and initiatives and fostering relationships and synergies between various groups of stakeholders (public, private, third sector and associations, citizens) and the different policy domains that food intersects (environment, production activities, logistics and transportation, education and training, economic development and employment, culture and tourism, health and social welfare) in a holistic and integrated vision [24].

Within this perspective, the citizen is assigned an active role, becoming a prosumer (from the crasis of producer and consumer) rather than a passive. By producing food, prosumers attempt to bridge the gap between production and consumption in cities as well as in rural communities.

Bringing agri-food production into housing (or vice versa) can thus include designing and implementing new suburban or peri-urban districts conceived of as laboratories for sustainable agricultural production, housing and social interaction, innovation, education, employment, among other things.

"ReGen Villages," a Stanford University spin-off firm envisioning the future of living in regenerative and resilient communities, has developed an innovative program-planning technique. This is a new visionary model for the establishment of integrated and resilient off-grid ecovillages that blend technology, innovation, circular economy, and self-sufficiency, including especially food. Positive energy housing, renewable energy production and storage, high-yield organic food production, aquaponic/aeroponic farming systems for vertical agriculture, water management, and waste-to-resource systems are among the innovative concepts embraced by the concept. In the new village concept, from the standpoint of a circular economy, the outputs of one system are actually the inputs of another. It also integrates artificial intelligence (AI) and machine learning (ML) to identify, create, and manage regenerative neighborhoods that promote long-term health outcomes for residents and communities. Moreover, these villages are planned for global replication and scale in collaboration with established industrial partners, universities, governments, and sovereign wealth and pension funds, enabling an optimistic green transition. Several architecture/engineering firms and companies in Europe have embraced this philosophy and collaborated with the US-based start-up to develop, propose, and experiment with new city and living models. According to White Arkitekter, Sweden could become the first country with circular and self-sufficient communities. Over the past few years, ReGen villages has met with several Swedish municipality administrations, landowners, real estate developers, and stakeholders with the aim of initiating, with the support of the Swedish architectural team, a pilot project in the country.

Naturbyen (Nature Village) is a similar experimentation (**Figure 1**) that was launched in Denmark in 2020 because of a shared desire among the municipality of Middelfart and a number of local communities to design an alternative future through a participatory process coordinated by the Danish design firm EFFEKT.

This collaboration led to the design of a housing area, conceived of as an international demonstration archetypal project of how sustainable housing development may be integrated with ambitious reforestation, improved biodiversity, and a circular approach to resources in suburban and peri-urban regions. Furthermore, housing contributes to agricultural output by creating healthy, socially integrated areas.

A total of 220 new residences located inside a newly planted forest outside big cities represent an alternative to the traditional terraced and parceled housing options that are still the most common housing typology in Denmark. The new municipality-led residential expansion project aims to become a laboratory for residential and agricultural development in suburban and peri-urban areas, with the goal of becoming an iterable intervention in similar realities and assisting Denmark in meeting its ambitious goal of covering 20% of its land area with forests by 2100.

A different approach is being pursued by the city of Shanghai. With a population of about 24 million people and a severe lack of agricultural land for food production, the Chinese megacity has envisioned a unique urban agricultural zone of roughly 100 ha.

The Sunqiao Urban Agricultural District, designed by Sasaki in collaboration with various stakeholders from both the public and private sectors, aims to meet the region's growing agricultural demand while also serving as a living, dynamic urban laboratory for research and innovation, social interaction, and education. In the new district, agriculture is introduced on a large scale through diffuse and punctual vertical farming interventions that take advantage of hydroponic and aquaponic cropping systems that have higher spatial-productive efficiency and a significant reduction in

DOI: http://dx.doi.org/10.5772/intechopen.113213

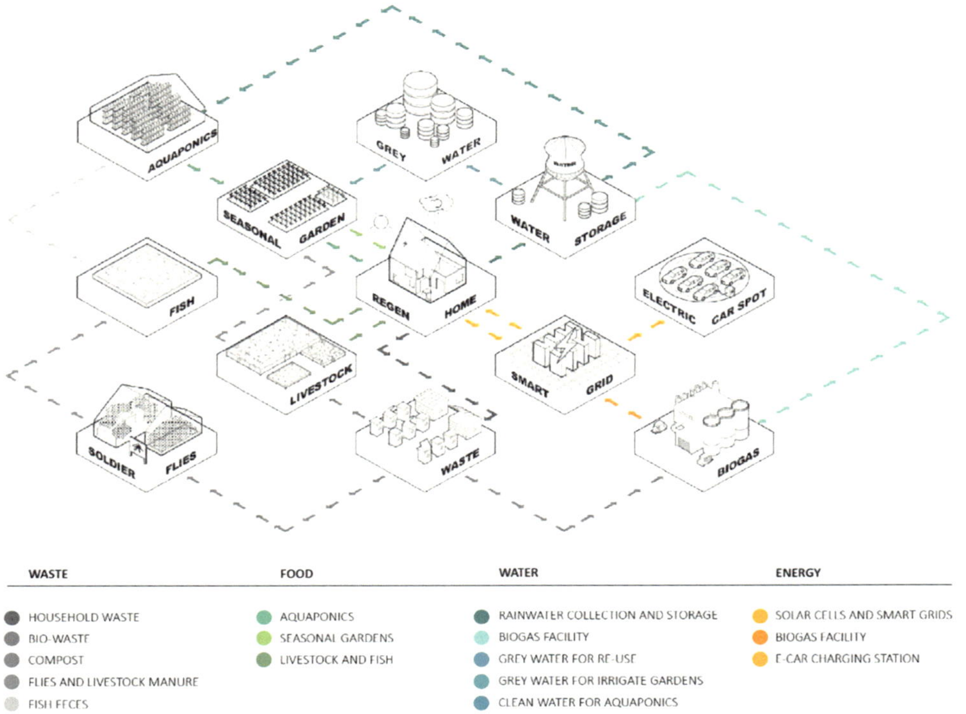

Figure 1.
Layout of an agri-city on the model of Naturbyen by Studio EFFEKT.

water and soil consumption. However, Sunqiao is more than just a food-production district. In fact, the intervention has a high social value and prioritizes agriculture as a key driver for urban growth. An interactive greenhouse, a science museum, and a market represent an attempt to educate future generations about conscious food consumption. Public spaces and facilities, offices, and houses represent the desire to create a mixed-use, dynamic, and active environment, far from the traditional concept of an agricultural district.

3. Floating potential for urban embedded food production. A zero-land footprint strategy

In recent decades, we have witnessed two related trends: land occupation on the one hand, and soil sealing on the other, both of which are the result of city growth and expansion of urban areas characterized by high building density ratios. Due to scarcity of available empty land within cities, agri-food production systems often play a marginal role in temporal (transient), spatial (interstitial), social (e.g., women and low-income groups), and economic (e.g., financial crisis, food shortage) terms. Indeed, urban farming and food production in cities are currently limited to the transformation of brownfields, residuals, and urban voids into micro-farming, private rooftop cultivation, urban community and institutional gardens, small-scale urban farming, urban aquaculture and aquaponics, urban forestry, and hydroponic and aeroponic vertical farming systems. In response to scarcity of land and/or water resources, the spread of vertical farming in cities has grown significantly, allowing for the combination of housing and farming within a single building inside the city.

Recently, a zero-land-footprint strategy that takes advantage of the continental and tidal hydrographic network for food production has gained popularity. This strategy entails the use of floating structures along rivers, lakes, or coastlines to house greenhouses and farms within urban centers.

Water proximity has always been a crucial component in the establishment and development of human settlements [30]. Many cities were built up along coastlines or at the mouths of large rivers because they served as collection points for raw materials coming from the inner areas, they were supplied by an efficient water transportation network and were guaranteed with access to clean water. Cities lacking permeable and underused soil but located near rivers, lakes, or coasts could easily host water-based food production facilities. The hydrographic network or the sea itself provides a huge potential for the floating development of food producing facilities in cities characterized by high building density. More precisely, floating farming facilities can provide several environmental and sustainability advantages [31], including: reducing the burden on freshwater by using seawater desalination techniques or collecting and storing rainwater; introducing new cultivable or breeding surfaces where permeable land and freshwater are scarce, particularly in high density urban areas; providing complete and self-sufficient farming systems in terms of automated planting, harvesting, processing, and export, drastically reducing transport costs; and providing the possibility of relocation in more appropriate sites when a given location is no longer suitable for any reason (environmental or pollution risks, political conflicts, and urban population shifts). Furthermore, floating greenhouses or breeding farms could be designed as multilevel vertical systems to increase overall farming surface and yield, ensuring the economic viability of the floating farm concept.

Floating agriculture is actually a vernacular soilless practice widely spread over Southeast Asia (Lake Inle Kay La floating village with farming and fishing arrangements), Middle East (Al-Tahla floating Islands in the southern wetlands of Iraq), and South America (Totora reed floating islands in Lake Titikaka, Peru). Different low-tech systems have been used for thousands of years and have allowed farmers to grow crops in flood-prone areas, wetlands, or lakes, where no other land use was conceivable. These systems usually consist of plants on rafts made of composted water weeds piled up on water bodies, by simply stripping nutrients released from decomposing organic material [32]. These systems are now seen as a strategy to cope with the combined effects of urbanization, land consumption, cementification, and climate change in areas that are more vulnerable to sea-level rise and coastal erosion, where flooding prevents land from being used for agriculture for extended periods of time [33] or where there is no available land for agri-production.

Floating Farm 2.0, designed and built by Goldsmith Studio, is the world's first floating dairy farm, located in the port of Rotterdam. The Floating Farm Dairy is a compact and efficiently stacked urban farm with a strong public and educational character. The building combines technical installations, storage, production, and processing of dairy on board. The farm produces fresh dairy products from its 40 cows. All raw dairy products are processed on-site and delivered across the city as fresh milk and yogurt. Floating Farm 2.0 is designed according to a circularity concept to employ leftover goods produced by the city, such as grass from public parks and food waste, to feed animals and return fresh milk to the city. This circular approach not only finds a new effective use for leftover products, but it also reduces food transportation costs and pollution by keeping food production and consumption tightly linked. Throughout a highly sustainable closed loop, cow manure is reused to produce fertilizer for public spaces within the city. The concept is envisioned for a future in which rising sea levels make farmland increasingly unusable due to flooding. The goal is to consider a new approach to bring agriculture back to the city while minimizing resource depletion and environmental impact and building resilience to climate change in a time-based design conception.

Another example of integrated agri-food production in the city is the Jellyfish Barge (**Figure 2**), a floating greenhouse module that aims to minimize energy, water, and soil footprint. Jellyfish Barge uses hydroponic cultivation with 70% water savings compared to traditional agriculture. The barge is made of recyclable materials and uses solar distillation to collect and purify 150 l of salt water per day. Fifteen percent of the seawater is returned to the water to improve the mineral content and nutritional value of the crops. One module is around 70 m^2 and can grow between 1400 and 1600 plants per month. One hectare could host more than 120 apartments.

Of course, these strategies do not expect to address the problem of city feeding by minimizing transportation and producing all consumed foods locally. Given the population figures, this scenario is inconceivable if only the continental hydrographic network is used as a new farming surface. But, as future visions suggest, one could even consider close offshore waters as farmland.

The Forward Thinking Architecture firm is branching out and transforming the way we think about agriculture and water. Its smart floating farm (SFF) concept is at the heart of this new way of thinking, and it is a real and already buildable construction. The floating farm is an offshore three-story floating facility that will host large hydroponic crops and fish farms beneath them. It is designed to be built off the coast of a city to produce both fish and vegetables using a simple system of linkages between different operational layers. The structure's composition is inspired by traditional Asian fish floating farms, but it also features two additional layers, one for growing any type of plant and another to supply the needed energy through solar energy conversion. Aside from the actual growth of plants (automated hydroponics) and hatching of fish, water-access points and a desalination plant (to convert ocean water to freshwater and then use it for farming) are provided, as well as an abattoir for the fish and a packaging facility. Solar panels, wind turbines, and wave energy converters have the potential to convert natural forces into useful electricity. It has the ability to produce 8.1 tons of fruits and vegetables and 1.7 tons of fish per year. The factory would be almost completely automated using sensor systems to capture data and fine-tune the

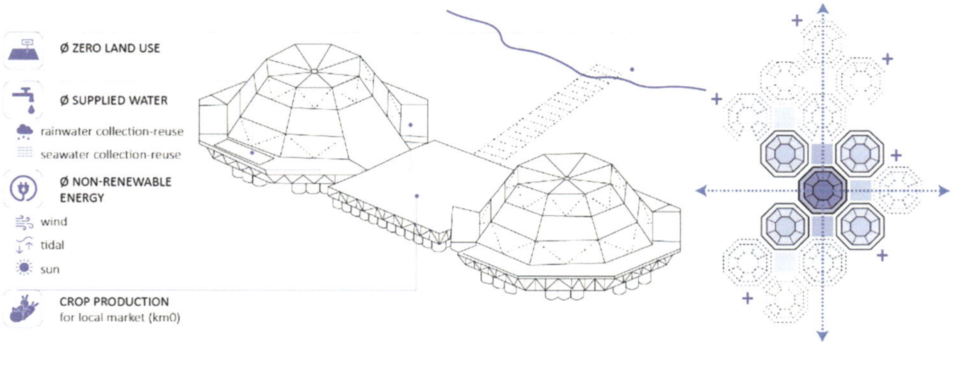

Figure 2.
Diagram of floating food production system on the example of Jellyfish Barge.

farms to work as effectively as possible. At the moment, it's an extremely ambitious concept. Yet, it raises a significant point: we could feed ourselves with low ongoing costs if we simply used endless and predictable resources such as the sun and the ocean.

Due to the high expense of desalination systems to produce irrigation water, and to the low salt tolerance of crops, alternative technologies have gradually emerged. Japanese start-up N-ARK has combined salt-tolerant technology with floating architecture to tackle the issues of sea-level rise and salt damage. In partnership with CULTIVERA agri-tech company, they aim to build a prototype of a floating marine farm "green ocean," conceived to float on the coast along urban areas. The facility makes use of a seawater agriculture technique based on moisculture, a humidity-controlled cultivation technology that reproduces the natural soil surface layer of about 15 cm using special fibers of 5 mm in diameter. Saline agriculture fertilizer is produced, thanks to a special circular process that absorbs water and nutrients in the air and mixes and neutralizes alkaline seawater and acidic rainwater. Moisculture requires only one tenth of the amount of water used in conventional irrigation farming.

The current challenge toward more resource-efficient cities is to shift cities metabolism from linear to circular, so that discarded material can become a resource for another process. Nutrients and carbon dioxide are two of the most common waste products generated by cities, and both are rarely reused or recycled before being discharged into the environment. A possible way to recycle nutrients and carbon dioxide is to use them as input for algae cultivation. Because of their ability to fix carbon via photosynthesis at up to 50 times the rate of terrestrial animals, algae are among the greatest organisms for CO_2 sequestration.

Cities are suitable locations for local recycling of waste due to their high concentration of nutrients and carbon dioxide. Unfortunately, dense urban areas often lack the space to implement large-scale algae cultivation. One alternative is to cultivate algae on the water, resulting in floating systems for biofuel and food production [34].

Another widespread practice is the integration of aquaculture within wider farming systems, contributing to the development of synergies between farming operators. Such systems are known as integrated agri-aquaculture systems (IAAS) and can help to improve water-nutrient balance through chemical or natural fertilization [35]. Agri-aquaculture systems generally comprise three major subsystems: aquaculture, agriculture, and household. Common positive interactions of agri-aquaculture systems include the use of animal manure as pond fertilizer, the use of crop by-products as supplementary feed for fish, the use of pond sediments as terrestrial crop fertilizers, and the use of aquaculture wastewater for crop irrigation.

Overall, producing, processing, and packaging food inside the city can significantly shorten the supply chain, add a certain social value and level of food security, contribute to new forms of urban circularity, and promote an efficient use of scarce space due to the lack of it. As a result, combining production facilities with the urban environment is expected to boost economic feasibility while also providing climate-proof expansion for a growing urban population.

Acknowledgements

The authors acknowledge the Italian Ministry of Education, University and Research (MIUR) for financial support. The essay is part of a MIUR-funded research project entitled "Public city and new welfare. Urban regeneration for the right to health"; PI: Prof. Alessandra Battisti; Protocol number: RM120172B3B15DB9.

Author details

Alessandra Battisti, Alberto Calenzo and Livia Calcagni*
Department of Planning, Design, and Technology of Architecture, Sapienza University of Rome, Italy

*Address all correspondence to: livia.calcagni@uniroma1.it

References

[1] World Bank. Poverty and Shared Prosperity 2022: Correcting Course. Washington: The World Bank; 2022. DOI: 10.1596/978-1-4648-1893-6

[2] United Nations. Department of Economic and Social Affairs, Population Division. World Population Ageing 2019 Highlights. New York: United Nations; 2019

[3] Waite R, Vennard D. Without changing diets, agriculture alone could produce enough emissions to surpass 1.5 C of global warming. 2018

[4] United Nations Food and Agriculture Organization. The State of Food and Agriculture 2021: Making Agrifood Systems more Resilient to Shocks and Stresses. FAO; 2021

[5] Benton TG, Bieg C, Harwatt H, Pudasaini R, Wellesley L. Food system impacts on biodiversity loss. In: Three Levers for Food System Transformation in Support of Nature. London: Chatham House; 2021

[6] Shukla PR, Skea J, Calvo Buendia E, Masson-Delmotte V, Pörtner HO, Roberts DC, et al. IPCC, 2019: Climate Change and Land: An IPCC special report on climate change, desertification, land degradation, sustainable land management, food security, and greenhouse gas fluxes in terrestrial ecosystems. 2019

[7] Gerber PJ, Steinfeld H, Henderson B, Mottet A, Opio C, Dijkman J, et al. Tackling Climate Change Through Livestock: A Global Assessment of Emissions and Mitigation Opportunities. Rome: Food and Agriculture Organization of the United Nations (FAO); 2013

[8] United Nations. World population prospects 2019: Highlights. In: Department of Economic and Social Affairs: Population Division. New York: United Nations; 2019. DOI: 10.18356/13bf5476-en

[9] Žičkienė S, Tamasauskiene Z. Social innovation for sustainable development. In: Innovations and Traditions for Sustainable Development. World Sustainability Series. Cham: Springer; 2021. pp. 47-68. DOI: 10.1007/978-3-030-78825-4_4

[10] Guo H, Liang D, Sun Z, Chen F, Wang X, Li J, et al. Measuring and evaluating SDG indicators with Big Earth Data. Scientific Bulletin. 2022;**67**(17):1792-1801

[11] Gupta S, Degbelo A. An empirical analysis of AI contributions to sustainable cities (SDG 11). In: The Ethics of Artificial Intelligence for the Sustainable Development Goals. Philosophical Studies Series. Vol. 152. Cham: Springer. pp. 461-484. DOI: 10.1007/978-3-031-21147-8_25

[12] United Nations General Assembly. Transforming Our World: The 2030 Agenda for Sustainable Development. New York: United Nations; 2015

[13] Cavalleri MC, Cournède B, Özsöğüt E. How responsive are housing markets in the OECD? National level estimates. OECD Economics Department Working Papers, No. 1589. Paris: OECD Publishing; 2019. DOI: 10.1787/4777e29a-en

[14] Cournède B, Ziemann V, De Pace F. The Future of Housing: Policy Scenarios. OECD Economics Department

Working Papers, No. 1624. Paris: OECD Publishing; 2020. DOI: 10.1787/0adf02cb-en

[15] Smets P, Kusenbach M. New research on housing and territorial stigma: Introduction to the thematic issue. Social Inclusion. 2020;**8**(1):1-7

[16] Marsden T, Sonnino R. Rural development and agri-food governance in Europe: Tracing the development of alternatives. In: Agricultural Governance. London: Routledge; 2007. pp. 66-84. DOI: 10.4324/9780203698907

[17] Donald B, Gertler M, Gray M, Lobao L. Re-regionalizing the food system? Cambridge Journal of Regions Economy and Society [Internet]. 2010;**3**(2):171-175. DOI: 10.1093/cjres/rsq020

[18] Wiskerke JSC. On places lost and places regained: Reflections on the alternative food geography and sustainable regional development. International Planning Studies [Internet]. 2009;**14**(4):369-387. DOI: 10.1080/13563471003642803

[19] Batzing W. Città e campagna al tempo della globalizzazione. Quali prospettive per gli spazi rurali in Europa? In: Le frontiere della geografia. Torino: UTET Università; 2009. pp. 61-68

[20] Marx K. Il capitale, libro III, Il processo complessivo di produzione capitalistica. Eugenio Sbardella, editor. Roma: Newton Compton Editori; 2016. VII sezione, I redditi e le loro fonti, cap, 48, 1

[21] Gandy M. Cities in deep time. City [Internet]. 2018;**22**(1):96-105. DOI: 10.1080/13604813.2018.1434289

[22] FAO (Food and Agricultural Organizations of the United Nations). Challenges of food and nutrition security, agriculture and ecosystem management in an urbanizing world. In: Food, Agriculture and Cities. Rome: FAO; 2011. p. 48

[23] Steel C. Hungry City: How Food Shapes Our Lives. London: Random House UK; 2013

[24] Dansero E, Pettenati G, Toldo A. La città e le filiere del cibo: verso politiche alimentari urbane. In: Quali filiere per un progetto metropolitano Slow tourism spazi comuni, città. Milano: FrancoAngeli; 2015. pp. 186-208

[25] Wiskerke JSC, Viljoen A. Sustainable urban food provisioning: Challenges for scientists, policymakers, planners and designers. In: Sustainable Food Planning: Evolving Theory and Practice. Wageningen: Wageningen Academic Publishers; 2012. pp. 19-35. DOI: 10.3920/978-90-8686-187-3

[26] Krzysztofowicz M, Rudkin JE, Winthagen V, Bock AK. Farmers of the Future. Luxembourg: Publications Office of the European Union; 2020. DOI: 10.2760/680650

[27] Bocchi S, Maggi M. Agroecologia, sistemi agro-alimentari locali sostenibili, nuovi equilibri campagna-città. Scienze del Territorio. 2014;**2**:95-100

[28] Viljoen A, Bohn K. Second Nature Urban Agriculture: Designing Productive Cities. New York: Routledge; 2014

[29] Sommariva E. Agricoltura Urbana strategie per la città dopo la crisi. In: Atti XV conferenza Nazionale Società Italiana Urbanisti-L'urbanistica che cambia rischi e valori Pescara. 2012. pp. 10-11

[30] EEA. L'acqua in città [Internet]. 2018 [cited 2023 Jun 14]. Available from: https://www.eea.europa.eu/

it/segnali/segnali-2018/articoli/l2019acqua-in-citta%0A

[31] Moustafa K. Ships as future floating farm systems? Plant Signaling & Behavior. 2018;**13**(4):e1237330

[32] Pantanella E, Cardarelli M, Danieli PP, MacNiven A, Colla G. Integrated aquaculture-floating agriculture: Is it a valid strategy to raise livelihood? In: XXVIII International Horticultural Congress on Science and Horticulture for People (IHC2010): International Symposium on 921. 2010. pp. 79-86

[33] Parvej H. Adapting to Waterlogging Situation Through Promotion of Floating Garden (Hydroponics) - Involving the Affected Community for their Sustainable Livelihood-Security. Proj report ActionAid. 2007

[34] Zanon BDB, Roeffen B, Czapiewska KM, de Graaf-Van Dinther RE, Mooij PR. Potential of floating production for delta and coastal cities. Journal of Cleaner Production. 2017;**151**:10-20

[35] Zajdband AD. Integrated agri-aquaculture systems. In: Genet Biofuels Local Farming Systems. Sustainable Agriculture Reviews. Vol. 7. Dordrecht: Springer; 2011. pp. 87-127. DOI: 10.1007/978-94-007-1521-9_4

Chapter 4

Mobile Housing as an Initial Proposal to Manage Informal Territories Exposed to Disaster Risks

Yasna Contreras and Beatriz Seguel

Abstract

From the study of informal territories, mobile housing is presented as an initial proposal to manage heterogeneous precarious settlements located in the north of Chile. More than a housing typology with certain construction characteristics, mobile housing contributes to a set of arguments that question housing policies and governmental reactions to urban informality, and its proximity to multiple human and physical hazards. In terms of management, the exposure of informal territories to numerous disaster risks has been a controversial aspect for local and central authorities, where the main actions have focused on eviction without temporary housing solutions, or the construction of supposed transitory neighborhoods, which are configured as territories of permanence. For this reason, mobile housing also raises the need to recognize that any existing and future housing must seek to reduce disaster risks, and must use materials according to the climatic, cultural, and temporal conditions of each territory.

Keywords: informal territories, urban informality, housing policy, disaster risk, disaster risk reduction

1. Introduction

Chilean housing policy has been known as a successful program [1]. Throughout its history, it has positioned itself as a solution to urban informality, being the main problem identified in Chilean cities since the beginning of the twentieth century [2]. From that moment on, the state was characterized by acting reactively to reduce the housing deficit and, in the same way, resolved the precarious living conditions of the people [2, 3]. However, contrary to the housing scheme developed through the first decades of the twentieth century, with the beginning of the military dictatorship in 1973, there was a turn toward a state with neoliberal foundations that understands housing as a commodity and a financial asset [4, 5]. This allowed privatization of social housing production and a consecutive promotion of housing solutions on the private market, meanwhile the state subsidizes the whole process and is responsible for selecting possible inhabitants [6, 7].

Despite this, the current construction of social housing has been reduced, while private housing units have tended to become more expensive, leading to a housing deficit calculated at around 641,000 units in 2022 [8, 9]. Its calculation includes the existence of housing in unlivable structural conditions, overcrowded homes, homeless people, and the growth of informal settlements [9]. The Chilean north is where the construction of informal settlements has had an accelerated and significant growth, concentrating 27% of settlements nationwide [10]. In this region, a complex contradiction is observed, since it is an extractive mining territory that contributes considerably to the economic growth of Chile and its gross domestic product [11]. Mining in connection with unregulated urban planning has promoted unequal and segregated cities, especially reflected in the value of land and access to housing for its population [12, 13] This is explained by the usual mobility of mining workers, whose origin from different parts of the country has driven processes of urban expansion due to real estate speculation [14, 15]. An example of this is the consolidated construction of high-standard housing in the most expensive areas of cities versus the emergence of informal territories, which includes the informal rental and sublease market in central spaces, in addition to informal settlements [12, 16, 17].

In this context, nearly 22,800 families have taken the action of occupying vacant land for the purpose of self-building homes in northern Chile [10]. Among them, there is Chilean population and immigrants from the Latin American and Caribbean region, who settle in peripheral areas within the cities. Although the informal territories can settle in any sector of the cities, their location in disaster-risk areas is recurrent [18, 19]. However, its informality condition has resulted in authorities excluding them from socio-natural disaster prevention programs, leading to an endless accumulation of exposure on its inhabitants [20, 21]. Bringing housing to the center of the disaster debate as reference [19] suggests, we propose mobile housing as an initial proposal to reduce disaster risks that affect informal territories, contributing with suggestions that go beyond the idea of just evicting. More than a housing typology with certain construction characteristics, mobile housing contributes with a set of arguments that question housing policies, governmental reaction to urban informality, and its view on people's residential trajectories. This proposal is based on long-term development of mixed methodologies, started in 2014, that has allowed us a comprehensive understanding of access to housing. It includes the analysis of quantitative data, such as demography, housing, overcrowding, land prices, purchase and sale prices of homes, and rental and sublet prices. Also includes the application of qualitative methods, such as 200 in-depth interviews with residents of the northern cities of Arica, Iquique, Alto Hospicio, Antofagasta, Calama, and San Pedro de Atacama, the development of housing evaluation sheets designed to investigate the architectonic background of informal housing, and the development of meetings and workshops with communities living in informal settlements to learn about their trajectories, views, and opinions.

2. Informal territories and Chilean housing policies

In Ref. [17] we define informal territories as places where different logics of access to land and housing coexist, recognizing a continuous interaction between the formal/informal and legal/illegal sectors. By proposing this concept, our intention is to continue discussions raised by theoreticians from the Latin American region on terms, such as informal settlements and urban informality. These concepts were widely discussed between the 1970s and 1980s when the welfare state was in crisis and many families could not access housing and land within the consolidated city. The informal settlement or urban informality moved from a legal-normative definition referring to spaces that conflict with current urban regulations, toward the conception of spaces not integrated into the conventional urban system [22–24]. In parallel, in the last 30 years, numerous investigations have exposed a new perspective where urban informality is referred to as a constituent part of the productive and territorial structure of cities [16, 25]. That is why informal territory concatenates the terms of territory and informality, exposing the sense of belonging and community that the population develops while seeking or developing residential alternatives in response to obsolete or limiting proposals provided by the state.

The concept of informal territory arises from urban phenomena seen in northern Chile, where space is dominated by socio-territorial injustice. In the northern cities, it is possible to find informal territories of different types, located in central, pericentral, and peripheral areas. This evidences their configuration as particular places since they have a heterogeneous condition regarding their creation, construction materials, housing purpose, and those who produce and inhabit them [17]. Likewise, informal territories imply the production of place in any geographic context [18]. However, a highly complex aspect lies in its location close to disaster risk areas, exposure to contamination, the presence of household waste, or in the case of informal settlements, its construction on old dumps. For this reason, we also postulate the informal territory as a socio-analytical and political category that recognizes various territorialities of informality in access to housing and land. It also observes the complexity of the households that produce it, as these spaces are not inhabited exclusively by families inserted in the classic notions of poverty, as was stated in the past. Rather, it is produced by households with reduced social mobility, households that have had to prioritize how to use their limited income (e.g., between paying the children's school fees or paying the rent for the home they occupy). Informal territories are also produced by immigrant households that, due to various situations, cannot access formal, well-located housing at a price according to their family income.

Therefore, informal territories also reflect the dynamic and heterogeneous forms of living that occur within the city, which can vary over time and can differ from one city to another [17]. As an exercise based on observations in northern Chile, we postulate four typologies of informal territories according to their level of consolidation, described in **Table 1**. Informal territories are configured in parallel as an alternative production to housing and as a survival mechanism for vulnerable and low-income groups. For this reason, they make visible different forms of agency, which are defined by the type of dwelling inhabited, the agreements they establish with the authorities, and their place of location in the cities. Following **Table 1**, first, we observe that the marginalized informal territories lack internal organization, being criminalized and racialized spaces, where the morphology of the occupied space complicates their possibility of articulating with other groups. Second, those who live in deteriorated informal territories cannot establish a community, since informal subletting avoids any type of political organization that would show the risks to which families are exposed, either inside the buildings or in their surroundings. Finally, in transitional and consolidated informal territories, there is evidence of a greater capacity for organization, whether in housing committees or cooperatives. Most of our interviewees declare that a consolidated informal territory gives them greater freedom to coordinate collectively, an essential factor for negotiating their self-managed urbanization projects.

Typology	Description	Community organization	Political agency	Geography - hazard exposure
Marginalized informal territories	Recently produced spaces. Precarious facilities. Produced by vulnerable homeless groups looking for survival housing.	No internal organization is given recent installation time.	No political agency.	Located in marginal spaces of the city, including the coastline, sloping land, and riverbeds. Exposed to tsunamis, floods, and landslides.
Deteriorated informal territories	Spaces produced by informal speculators who sublease deteriorated homes. Precarious infrastructure with overcrowding, health problems, and insecurity.	Space without internal organization is given the control and rules established by informal landlords.	No political agency.	Located within the consolidated city. Exposed to the fire hazard due to overcrowding and overuse of electricity.
Informal territories in transition	Spaces produced by the state as a solution to informality and a transitory step toward permanent housing.	Housing committees or other similar organizations.	Organized for settlement, or to be relocated to permanent housing.	Located on the peripheries near riverbeds and slope sectors. Exposed to flooding, landslides, and pollution.
Consolidated informal territories	Long-standing spaces with high population density. They incorporate political projects that seek settlement and self-management.		Organizations that actively participate in settlement negotiations.	
Source. Authors.				

Table 1.
Typologies of informal territories in northern Chile.

On the link between informal territories and public policies, since the 1990s there has been a gradual redirection toward policies that seek to overcome what is called the problem of informal settlements [26]. The institution in charge of developing these policies has been the Ministry of Housing and Urbanism, together with all its associated services, which have developed various programs to settle or eradicate informal settlements by placing their inhabitants in social housing complexes [26, 27]. Here people must apply for housing subsidies, being an economic aid to finance access to social homes, considered the definitive housing solution for people and their families. Unfortunately, the relationship between inhabitants of informal settlements and housing policies has had some historical and current frictions. When a group has decided to leave informal settlements and relocate to social housing complexes, people would encounter hostile urban environments, usually marked by violence, leading them to question how owning these houses would grant them the dignity and adequate living conditions assured by the government and its authorities [28, 29]. The peripheral location of the land available to build social housing has registered important consequences in the displaced groups, including lack of connectivity and transportation to access workplaces, education, and health establishments, in addition to the loss of support networks, leading in many cases to people wanting to return to informal settlements [29]. This allows us to reflect on how individual residential trajectories are conceived by the state, and if access to the property would be the last stage of it. Finally, we do not mention subleased dwellings in central spaces, since there is a lack of programs for their inhabitants. A preliminary hypothesis would explain that the occurrence of housing violations within the urban limits and behind formal housing facades, would make it difficult to identify this situation and with that, to develop public policies to solve it.

2.1 Specific housing policies in northern Chile

In addition to evictions due to irregular occupation of land, it is common for northern city authorities to promote the relocation of informal settlements due to their current location in risk areas, such as riverbeds and sloping areas, all of which are susceptible to landslides and floods. There are also informal settlements located in proximity to critical but high-risk infrastructures, such as high-tension electrical towers and sewage processing plants. Considering that there is a high and growing population residing in informal settlements, authorities recognize the impossibility of evicting without having housing solutions. For this reason, five years ago, a program called transitory neighborhoods was developed. These are territorial units promoted by the state located on public land, which maintain the physical infrastructure of informal settlements, characterized by houses made of light material. Public officials fulfill the role of supervising who can access these neighborhoods, organizing the location of houses and equipment, and mediating conflicts between residents, in addition to managing legal access to basic services, such as electricity and drinking water, something impossible in standard informal settlements.

Authorities explain two purposes for the transitory neighborhoods, meanwhile, it is a program still on a exploratory stage. First, there are neighborhoods seeking to establish themselves as formal housing in the future. For this, a possible purchase of the occupied land is being managed, in addition to contemplating the deployment of housing programs. Second, other neighborhoods are installed as temporary residence spaces for people who want to leave informal settlements and access social housing. People grouped in housing committees circulate here, waiting in the temporary neighborhood until their respective housing complex is built. This contemplates the considerable passage of groupings. For example, our findings record the passage of six housing committees in a single transitory neighborhood in the city of Antofagasta. Although it has been proposed at one time as the solution to the "problem" of informal settlements, transitory neighborhoods maintain the same logic of subsidies and historical housing programs, where to resolve informality, one must access home ownership, without considering that residents may want to explore other forms of tenure. Even for authorities, signing up for a housing committee would be a dead end in terms of residential trajectory. "What are you doing within a housing committee? Because with this you will be tied down for the rest of your life," declared an authority from the northern regional housing service. Rental subsidies can also be used by those who wish to move out of informal settlements. However, it is a program that does not consider temporary housing infrastructure to await the results of the application. It is necessary to advance toward providing housing solutions from the idea of shelter, specially for those who live in constant precariousness and under the possibility to remain homeless, in addition to those who do not plan to settle in a particular city.

2.2 Housing policies and real estate as drivers of disaster risk

While we have referred to hazard exposure in informal territories, our findings in northern cities show that there is formal housing for high-income groups also settled in areas with tsunamis, floods, or landslides hazards. However, contrary to the situation of the informal territories, it is common for the exposure of high-income groups to be accepted, as they would have sufficient resources to face and overcome a disaster. This would explain the constant construction of private housing in contraindicated urban sectors, although it is worth wondering if people with lower incomes are actually living in safe areas. Some authors have identified the tendency in northern Chile to build social housing near risk areas [12, 30, 31]. Yet we would like to expose a particular example observed in the city of Antofagasta where informal territories, the construction of social housing, and the risk of disaster overlap.

After Santiago, Antofagasta is the second most expensive city to reside in Chile. On its surface, there are different types of housing with differentiated values according to square meters and number of rooms. Given the difficulties of buying a home due to its high value, renting is consolidated as one of the main forms of tenure, however, its monthly values can reach USD $630 for a 45 m^2 apartment, meanwhile, renting a 3-bedroom house can cost USD $1056. This is also exacerbated by the social geography of the city, where rental values change according to location and accessibility to goods and services. As a result of these restrictions on access to housing for ownership, rent, or even sublease, some families see in peripheries an opportunity to self-build their houses. In terms of disaster risk, peripheral informal settlements are exposed to countless hazards given their location in the western part of the coastal mountain range, an area determined by a steep slope that is highly susceptible to floods and landslides [32]. Furthermore, any house that is built in this area must consider historical disaster events. The last significant episode was a mudflow that affected the whole city of Antofagasta in 1991, triggered by an intense rainfall that activated dry creeks, registering 92 fatalities and 16 missing people [32]. Despite this, these are areas where, in addition to informal settlements, the state has built social housing for vulnerable groups. There is a remarkable example of Altos del Arenal building, a social housing project developed in 2018.

Located in the northern area of the city of Antofagasta, Altos del Arenal is a social housing complex evaluated and approved by the regional housing and urbanization service (known as SERVIU), which allowed the construction of 50 apartments of 62 m^2, producing a six-floor tall building. Prior to its construction, it was defined that its residents would be some families living in Los Arenales, one of the largest informal settlements in the city. Candidates for relocation had to register in the overcoming camps program of the Ministry of Housing and Urbanism and apply for a housing subsidy to buy the apartments. However, as **Figures 1** and **2** show, the social housing complex was placed in the rear sector of Los Arenales, an informal settlement under

Figure 1.
Altos del Arenal social housing complex, Antofagasta. Source. Authors.

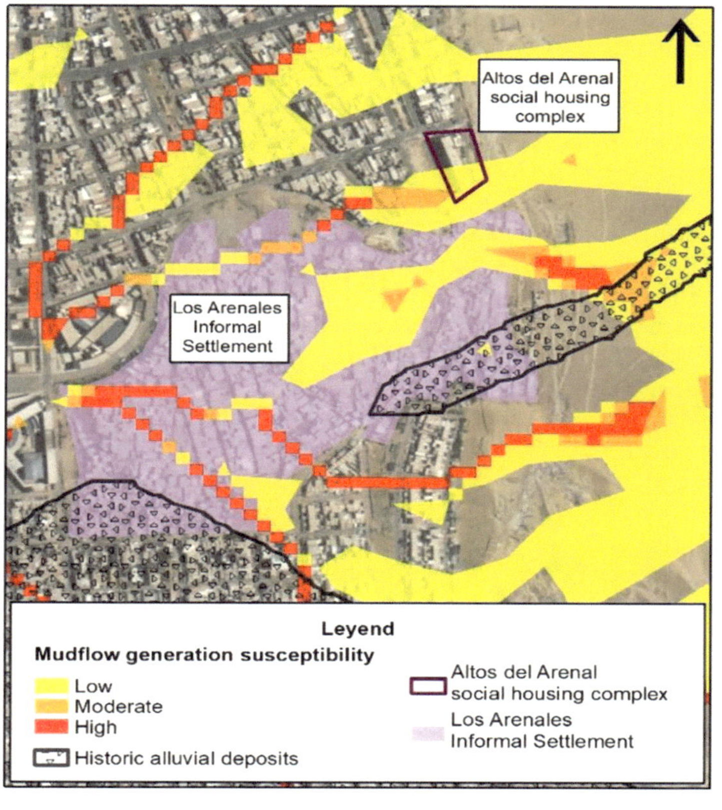

Figure 2.
Susceptibility to mass removal, altos del Arenal social housing complex, Antofagasta. Source. Authors.

constant risk of eviction. Likewise, it is an area under high susceptibility to landslides, in addition to historical mudflow records.

From this, state's decision to build social housing in risk areas, **Figures 1** and **2** above, expose some challenges to face. First, it is necessary to rethink the scale of disaster risk and its link with territorial planning instruments. In Antofagasta, the current urban regulatory plan was created 20 years ago, requiring it to be updated especially in risk issues, since zoning depends on hazard exposure studies, which have not been properly developed. And second, it is also necessary to investigate the exposure to anthropic hazards, such as proximity to dumps, the presence of pests, and pollutants, among others, as an exercise to begin unraveling the sociopolitical construction of risk in informal territories.

3. Mobile housing as an initial proposal and strategy to reduce risk

From the analysis of northern informal territories and the experiences of their inhabitants, the mobile housing proposal arises. More than a housing typology with certain construction characteristics, we propose mobile housing as a government housing program for cases of residential vulnerability, including those who have been evicted or displaced, to leave informality given their situation of helplessness that can directly lead to homelessness. Specifically, mobile housing would answer an intermediate stage of the residential trajectories of individuals, located between the complete housing precariousness, common in some informal territories, and access to permanent housing.

This program also would allow access to formal housing under a provisional scheme that recognizes their possibility of choosing where to reside for a certain time without leading directly to own a property, in addition to being able to decide how to continue their residential trajectory. The most explicit example of this situation would be the residential choices of immigrants, who could be temporarily living in a northern city, and then continue their journey to other latitudes if they wish so. In this sense, mobile housing would also make it possible to avoid highly detrimental conditions for well-being, such as overcrowding, lack of privacy, exposure to precarious living conditions, including construction with inappropriate materials that generates concentration of humidity and lack of natural light, and the experience of various types of violence or aggression, all especially present in the sublease of rooms in central spaces.

In practical terms, mobile housing would allow renting available housing in the built environment of urban or rural areas, considering the extensive difficulties exposed by the state to build social housing. This housing program should have more flexible beneficiary selection criteria than the current rental subsidy, contemplating a reduced time to be applied, while it seeks to avoid homelessness. Likewise, the program should consider that the beneficiaries reside in rented houses for a range between 6 months and 1 year. In terms of management, the program could contemplate a monitoring and orientation process by public officials that allow individuals and their families to make the best decisions about their residential trajectories and their future actions. With this, the mobile housing program would incorporate nomadism as a process of constant re-territorialization, as it allows negotiation between the new context where people are about to settle residentially and their previous experience in other places where they have lived [33].

At the same time, recognizing the significant amount of population that currently resides in informal territories, the mobile housing program would also consider improving the risk exposure conditions of these spaces. This line of intervention in

Dimension	Proposals	Change's agents
Education and risk governance	Workshops to raise awareness about the types of hazards to which each informal territory is exposed. Training to deal with emergencies. Define protocols and evacuation routes. Activation of emergency protocols from state services.	Communities Regional and/or local government. Emergency office and related institutions.
Control of habitability conditions	Regulation and inspection of central subleased properties. Evaluation of building conditions in informal settlements. Requirement to build firewalls. Check availability and maintenance status of fire hydrants. Garbage cleaning in creeks. Control of informal settlement expansion.	Regional and/or local government.
Mitigation infrastructure	Hillside afforestation. Construction of tree-lined furrows around houses and on edges of creeks. Transverse orchards in valley and foothill areas.	Communities. Regional and/or local government. Emergency office and related institutions. Environment and planning offices.

Source. Authors.

Table 2.
Proposals to reduce risk in informal territories.

informal territories does not necessarily seek the formalization or immediate regulation of informal settlements and deteriorated housing, but instead acknowledges that their inhabitants live at imminent risk and a planning response must be given. As a result, we suggest a series of proposals that cover various dimensions of living, reacting, and preventing risk, as shown in **Table 2**.

Finally, with these preventive steps, the constant absence of the state in informal territories would be compensated, since the condition of informality should not exclude its population from receiving adequate protection against risks. Likewise, historical disasters registered in northern cities show that any event affects formal and informal spaces transversally. Consequently, it is time for a territorial planning that projects adequate and safe conditions for all inhabitants.

4. Conclusion

From our research experience in informal territories, we propose a housing program for the Chilean government system that collects differentiated residential demands and allows vulnerable population to leave housing informality. In its range of areas, mobile housing recognizes the factors that drive the creation of informal territories, including the restrictive prices to access the private housing market, the difficulties faced by the state in producing social housing, in addition to the diversity of households that cannot afford none of the above options. From the geographical conditions where massive informal territories are settled, mobile housing seeks for inhabitants to live in safe conditions against disaster risks, even temporarily. This implies rethinking how the risk of disaster is reduced in homes located under certain climatic, edaphic, and arboreal conditions. It is crucial to propose how to build houses resistant to intense rains and humidity conditions, and to water scarcity and intense solar exposure. With this, mobile housing requires generating soil mechanics in seismic risk areas, while in areas with tsunami risk, establish the obligation to build high-rise homes where the habitable sections begin on the second floors. As noted above, about the risk of mudflow, mobile housing proposes the construction of furrows that surround the houses and water sinks, in addition to afforestation to stabilize the soil on slopes. Therefore, mobile housing and its proposals seek to configure a housing typology that can be replicated in any territory where institutions are expected to reduce disaster risk. Consequently, mobile housing also requires the Chilean state to stop actions that perpetuate the exposure to risks of vulnerable families, who after experiencing disasters are moved to new homes exposed to other hazards[1].

To conclude, in addition to being a housing program, mobile housing is also a theoretical-methodological concept that conceives housing as an object and subject, but which is inserted in territories with particular conditions. This multi-scale perspective highlights that inhabiting a space is not limited to housing as a productive and reproductive place, but also involves its surroundings. That explains the multiple proposals presented, which, above all, would favor living under the basic conditions of comfort that adequate housing requires. With this, we consider the future potential of mobile housing to reduce the housing deficit in all its stages, for example, allowing the state to effectively have land reserves, proposing resilient and adaptable housing models, in addition to the existence of risk mitigation subsidies that improve constructions without discriminating informal settlements or slums. Finally, mobile housing is an initial and exploratory exercise that also challenges the development of housing policies from an interdisciplinary perspective with the purpose of finding integral solutions.

Notes/thanks/other declarations

The authors would like to thank the different residents of informal territories interviewed and consulted over the years about their experiences and points of view. Also, they would like to thank the Directorate of Research and Creation of the Faculty of Architecture and Urbanism of the University of Chile for their support. The authors would also like to thank the Program to Support Academic Productivity in Social Sciences, Humanities, Arts and Education of the Vice-Rector's Office for Research and Development of the University of Chile N° PROA013/22.

Acknowledgements

This chapter includes findings from the following research projects:

ANID's National Fund for Scientific and Technological Development (FONDECYT Regular) N°1171722: "Geographies of access to housing for Latin

[1] For example, this situation was seen during the reconstruction after the 8.8 Mw earthquake that hit central-southern Chile in February 2010. The population affected by the subsequent tsunami was relocated to social housing built in areas at risk of forest fires and mudflow. [34].

American and Caribbean immigrants: Exploring new socio-spatial phenomena in cities of northern Chile."

Center for studies on conflict and social cohesion (COES - ANID FONDAP N° 15130009). Mini COES Project: "Immigrant-migrant population in informal territories and socio-environmental injustices: The case of the cities of Arica, Iquique-Alto Hospicio and Antofagasta."

Conflict of interest

The authors declare no conflict of interest.

Author details

Yasna Contreras* and Beatriz Seguel
University of Chile, Santiago, Chile

*Address all correspondence to: ycontrerasg@uchilefau.cl

References

[1] Ducci ME. Chile: el lado obscuro de una política de vivienda exitosa. Revista EURE - Revista de Estudios Urbano Regionales. 1997;**23**(69):99-115

[2] De Ramón A. La población informal. Poblamiento de la periferia de Santiago de Chile 1920-1970. Revista EURE - Revista de Estudios Urbano Regionales. 1990;**16**(50):5-17

[3] Hidalgo R. Vivienda social y espacio urbano en Santiago de Chile. Una mirada retrospectiva a la acción del Estado en las primeras décadas del siglo XX. Revista EURE - Revista de Estudios Urbano Regionales. 2002;**28**(83):83-106

[4] Gilbert A. Power, ideology and the Washington consensus: The development and spread of Chilean housing policy. Housing Studies. 2002;**17**(2):305-324

[5] Rolnik R. La guerra de los lugares. In: La colonización de la tierra y la vivienda en la era de las finanzas. Santiago de Chile: LOM Editores; 2017

[6] Imilan W, Olivera P, Beswick J. Acceso a la vivienda en tiempos neoliberales: Un análisis comparativo de los efectos e impactos de la neoliberalización en las ciudades de Santiago. México y Londres. Revista INVI. 2016;**31**(88):163-190

[7] Sugranyes A, Jiménez F. El derecho humano a una vivienda adecuada en Chile. In: Rodríguez A, Rodríguez P, Sugranyes A, editors. CON SUBSIDIO, SIN DERECHO. La situación del derecho a una vivienda adecuada en Chile. Santiago de Chile: Ediciones SUR; 2015. pp. 17-44

[8] Rodríguez A, Sugranyes A. Aportes para un debate. In: Rodríguez A, Sugranyes A, editors. LOS CON TECHO. Un desafío para la política de vivienda social. Santiago de Chile: Ediciones SUR; 2005. pp. 13-20

[9] Cero D. Déficit habitacional: ¿Cuántas familias necesitan una vivienda y en qué territorios? Boletín 1: Estimación y caracterización del déficit habitacional en Chile abril 2022. Santiago: Centro UC Políticas Públicas and Déficit Cero; 2022

[10] Techo - Chile. Catastro Nacional de Campamentos 2020-2022. Santiago: Informe ejecutivo; 2022

[11] Banco Central de Chile. Base de datos estadístico. Santiago, Chile: Producto Interno Bruto; 2022. Available from: https://si3.bcentral.cl/Siete/ES/Siete/Cuadro/CAP_CCNN/MN_CCNN76/CCNN2018_PIB_REGIONAL/637899744282781812

[12] Contreras Y, Neville L, González R. In-formality in access to housing for Latin American migrants: A case study of an intermediate Chilean city. International Journal of Housing Policy. 2019;**19**(3):411-435

[13] Vergara F. Expoliación de la vivienda como activo financiero de renta fija en Antofagasta. Antofagasta, Chile: Serie de Documentos de Trabajo en Economía; 2017

[14] Aroca P, Atienza M. La conmutación regional en Chile y su impacto en la Región de Antofagasta. EURE (Santiago). 2008;**42**(102):97-120

[15] Thodes E. Segregación socioespacial en ciudades mineras: el caso de Antofagasta, Chile. Notas de Población. 2016;**102**:203-227

[16] Contreras Y, Ala-Louko V, Labbé G. Acceso exclusionario y racista a la

vivienda formal e informal en las áreas centrales de Santiago e Iquique. Polis. 2015;**14**(42):53-78

[17] Contreras Y, Seguel B, Territorio informal. Una nueva lectura del acceso a la vivienda y al suelo en Chile. Revista de geografía Norte Grande. 2022;**81**:113-136

[18] Caldeira T. Peripheral urbanization: Autoconstruction, transversal logics, and politics in cities of the global south. Environment and Planning D: Society and Space. 2016;**35**(1):3-20

[19] Sandoval V, Sarmiento JP. Una mirada desde la gobernanza del riesgo y la resiliencia urbana en América Latina y el Caribe: Los asentamientos informales en la Nueva Agenda Urbana. Revista de Estudios Latinoamericanos sobre Reducción del Riesgo de Desastres (REDER). 2018;**2**(1):38-52

[20] Romero Toledo H, Romero AH. Ecología política de los desastres: vulnerabilidad, exclusión socio-territorial y erupciones volcánicas en la patagonia chilena. Magallania (Punta Arenas). 2015;**43**(3):7-26

[21] Allen A, Belkow T, Escalante C, De los Ríos S, Kamiya M, Lambert R, et al. De la mitigación de desastres a la interrupción de trampas de riesgo: La experiencia de aprendizaje-acción de cLIMA sin Riesgo. Revista de Estudios Latinoamericanos sobre Reducción del Riesgo de Desastres (REDER). 2017;**1**(1):6-28

[22] Azuela A. Los asentamientos populares y el orden jurídico en la urbanización popular en América Latina. Sociológica. 1993;**5**:12-17

[23] Duhau E. Dimensiones sociopolíticas de la irregularidad y la regularización de los asentamientos populares. Cambridge, MA: Lincoln Institute of Land and Policy. 2002; 57(1).

[24] Clichesvsky N. Pobreza y acceso al suelo urbano. In: Algunas interrogantes sobre las políticas de regularización en América Latina. Santiago de Chile: División de Desarrollo Sostenible y asentamientos humanos. Santiago de Chile: División de Desarrollo Sostenible y asentamientos humanos. CEPAL; 2003

[25] Herzer H, Di Virgilio MM, Rodríguez MC, Redondo A. Informalidad o Informalidades? Hábitat popular e informalidades urbanas en áreas urbanas consolidadas (Ciudad de Buenos Aires). PAMPA. 2008;**1**(4):85-112

[26] Matus C, Ramoneda Á, Valenzuela F. La integración social como desafío: análisis del programa de campamentos en Chile (2011-2018). Revista INVI. 2019;**34**(97):49-78

[27] Matus C, Ramoneda Á, Valenzuela F, Ruiz-Tagle J. Los campamentos en la política urbana chilena: desafíos y propuestas para abordar una problemática en crecimiento. Santiago: Centro de políticas públicas UC; 2020

[28] Murphy E. For a Proper Home: Housing Rights in the Margins of Urban Chile, 1960-2010. Pittsburgh: University of Pittsburgh Press; 2015

[29] Sugranyes A. Capítulo 1: La política habitacional en Chile, 1980-2000. In: Rodríguez A, Sugranyes A, editors. Los con Techo. Un desafío para la política de vivienda social. Santiago de Chile: Ediciones SUR; 2005. pp. 23-58

[30] Castro CP, Magallanes V, Soto MV. Riesgos socionaturales asociados a amenazas hidromorfológicas en la ciudad de Copiapó. In: Vargas G, Pérez S, Aldunce P, editors. Aluviones y resiliencia

en Atacama. Construyendo saberes sobre riesgos y desastres. Santiago: Social-Ediciones; 2018. pp. 135-148

[31] Hidalgo R, González M, Vergara C, Alvarado V. Vivienda y urbanización en la ciudad de Arica: costa, desierto e integración subsidiaria en la frontera norte de Chile (2000-2019). Diálogo andino. 2021;**64**:225-242

[32] Vargas G, Ortlieb L, Rutllant J. Aluviones históricos en Antofagasta y su relación con eventos El Niño/Oscilación del Sur. Revista geológica de Chile. 2000;**27**(2):157-176

[33] Silva L. Informalidad, Movimiento y Nomadalidad: Cinco tesis sobre informalidad urbana. In: Taller de proyecto de tesis 5. Programa Doctorado en Territorio: Espacio y Sociedad, Universidad de Chile; 2021

[34] Contreras Y, Arriagada C. Reconstrucción exclusionaria. Lo comunitario y las políticas públicas en ciudades menores e intermedias chilenas afectadas por el terremoto y tsunami del 27F 2010. Vol. 64. Los casos de Constitución y Dichato. Santiago, Chile: Revista de Geografía Norte Grande; 2016. pp. 83-107

Chapter 5

Morphological Evolution of Single-Family Dwellings: A Prospective towards 2100

Marco Avila-Calle, Jorge Toledo-Toledo and Federico Córdova-González

Abstract

The objective of the research is to project the morphological evolution of single-family dwellings towards the year 2100, using the scenario methodology for prospective studies, to obtain formal design strategies that can be adapted for a probable future. The research approach is qualitative documentary, using primary and secondary sources to obtain retrospective data and then can project them into the future. The methodological process is divided into six stages; in the first stage called planning, through bibliographic and audio-visual documentary analysis, four probable futures are established and studied by articulating timelines using retrospective and prospective data; in the second stage called anticipation, the most relevant historical milestones are identified to outline these futures; in the third stage, called temporisation, the formal patterns are studied through the redesign of the facades of the selected case studies; in the fourth stage, called deepening, the possible causes of futures that could develop are studied in depth and verified; in the fifth stage, the scenarios with the greatest plausibility and scientific evidence are presented; and in the sixth stage, the prospective scenarios are proposed to the experts to be validated by consensus. The results obtained with this research are a set of prospective scenarios of the morphological evolution of single-family dwellings and their formal design strategies.

Keywords: morphology, housing of the future, chaos, Gaia, evolution, progress

1. Introduction

> "Whatever a man can imagine, another man can make it come true"
>
> Jules Verne [1]

Two centuries ago, French writer Jules Verne, using his analytical and synthetic thinking, managed to describe the incipient technological advances of his time, projecting them a century ahead through his futuristic stories. When there were hardly any prototypes of submarines, helicopters or space rockets, Verne narrated his stories of journeys into outer space or to the bottom of the sea, prospective scenarios

IntechOpen

that by the end of the nineteenth century were becoming reality. One of his famous phrases was: whatever a man can imagine, another man can make it come true. This background allows us to assert that if we want to project our future, we must know our past and analyse our present. For as Prof. Sohail Inayatullah defines it in his proposal of the Triangle of Futures, there are different driving or inhibiting forces of change that emerges from each of the dimensions of time: the weight of the past, the push of the present and the pull of the imagined future. In this approach, alternative futures emerge in a whirlpool that forms between its three vertices. These vectors push and pull in different directions, each with its own set of driving and inhibiting forces. It is the balance between the force vectors at each corner that will define the different plausible futures [2].

In view of this, if we were to place the time horizon up to 2100, the following research questions arise: what will the houses of the future be like? can we glimpse the future of architecture? would it be plausible to predict future events? can we control the variables implicit in the design and construction of houses in the years to come? These are questions that are addressed in this academic study from a scientific point of view. However, the mere fact of talking about architecture entails a complete and complex set of knowledge, which is why in this academic analysis the study focuses only on the morphological production of architectural form, using as a case study the single-family dwelling in prospective scenarios.

1.1 Problem

> "If you kill a cockroach you are a hero, but if you kill a beautiful butterfly you are a bad guy. Morality has aesthetic criteria"
>
> Frederich Nietzsche [3]

Throughout the history of mankind, there have been great cultures that have excelled in creating architectural works that endure to this day, and these works have been catalogued of great architectural value for multiple characteristics; for example, in the ancient and middle ages, the architectural works of relevance are found around the temples of the gods, creating monumental works as an icon of respect for the divinity whose architectural value was represented by the beauty of its columns, friezes, capitals, arches, lintels, etc., with philosophical, religious or social principles that were well founded in each period [4]. With the arrival of the Renaissance, in the modern age, the human being focuses his attention on man as the centre of the world; however, the architectural production continues to consider the design parameters of the preceding epochs, and the formal value was still manifested in the beauty and monumentality of its façades. With the rationalist vision and the first industrial revolution, the contemporary era began, considered a turning point in the history of mankind, because, despite achieving an indisputable technological and economic development, there was also a social gap between developed and developing countries, as well as the beginning of global warming due to the industrial use of coal. In the architectural field, the production of works and their aesthetic value are related to the use of the materials of the time and their monumentality is reduced to the scale of man. Currently, architectural production revolves around the capitalist system, and the architectural work is treated only as a good or a commercial service that generates profits, without considering the philosophical principles with which it was founded [5]; however with climate change, globalisation, technological development and

DOI: http://dx.doi.org/10.5772/intechopen.108521

economic and social inequality, makes us think that architecture is in an experimental stage trying to adapt and project itself into new futures, where the beauty and value of the architectural work should be studied with a multi-criteria vision. Therefore, this research addresses a historical analysis of the formal production of architecture and its projection into the future, with a view to finding methodological and design strategies based on prospective scenarios.

1.2 Reference framework

> "Life is a series of collisions with the future; it is not the sum of what we have been, but what we long to be"
>
> José Ortega y Gasset [6]

As mentioned in previous paragraphs, to talk about an architectural work in an integral way entails having a complete and complex maturity and knowledge in art and science, considering that this study bases its analysis only on the aesthetic part of the architectural work, to be more precise on the morphology of the façades of single-family houses in prospective scenarios, it is important to specify some concepts that allow us to understand the context and discuss the results obtained.

1.2.1 Morphology

What is morphology? Morphogenesis is a term coined by biology to explain the evolution of form, especially in plants, so that in architecture the process of change, evolution, transformation or mutation of form is a critical point in the training of architects, since it is here that the future professional must develop and argue the creative process of generation and transformation of form so that the final product is articulated in a theoretical and methodological support where the relationship between form and discourse can be appreciated [7]. Why is it important to study morphology? Man and architecture today is a product of globalisation and technology, facts that have allowed the evolution of construction, but have also generated the loss of architectural identity of different cultures and this is reflected in current production globally, so it is important to understand the new role and place of the subject in the modern world and architectural process. For example, Y S Yankovskaya mentions that;

> "in Russia, as in the rest of the world, the morphological production of architecture has been related to the development of ideas of form under the influence of various philosophical and psychological concepts related to the perceiving individual and has progressed from concepts of visual perception by an abstract individual to those of the interpreting individual who humanises the architectural space with his or her presence; from the emphasis on the role of the subject/interpreter of language that assigns meanings to the material object, to the priority of the subject/client in architecture with its intersubjective interactions in the design process. Given this, the main role shifts to the specificity of interaction between the images that exist in the minds of various subjects/participants of the design process with respect to the future architectural object. These images reflect, with varying degrees of divergence, the set of ideas about the functions, structure, comfort, identity, and socio-cultural significance of the architectural object. It is therefore increasingly important to ensure that these user-preferred images are part of the morphological structure of the architectural object. Herein lies the importance of

> morphological study as there is a clear link between the image as a carrier of meanings and the morphology of the architectural object as the most significant direction of development for contemporary architectural theory" [5].

Who studies morphology in architecture today? The study of morphology in architecture is limited, very few studies are directly linked to the present research; for example, Yankovskaya [5], Fernando Fraenza [8], Clifford Geertz [9], Patricio Aguirre [10], Guillermo López [11], among others study morphology from a current or retrospective perspective; however, there are no validated studies of morphological prospective in architecture, which is where the value of the present research proposal lies.

1.2.2 Foresight scenarios

What are prospective scenarios? There are different ways of approaching the study of the future, for example astrology, prophecy, prognosis, prediction, forecasting, projection and prospective, the latter being the only one that approaches it from a scientific point of view, a multiple and indeterminate reality, obtained by the infinite possibilities of human action, reflected in the different projects, desires and fears of social groups, these are called prospective scenarios [12].

How are prospective scenarios proposed? This methodology analyses the phenomenon under study from a retrospective and current point of view, considering the influence of the social groups responsible for its development, to subsequently present the future reality in the form of scenarios [12], generally developed in three main phases:

- Structural analysis: In this phase, the past and present are studied in search of patterns that project future scenarios.

- Analysis of the actors' game: This allows for the identification of the main social actors, as well as the spatial and temporal context in which they operate, which makes it possible to identify the existing relationships between each human action over time, thus promoting the creation of different prospective scenarios.

- Scenario elaboration: Using the information collected in the previous phases, each researcher interweaves scientific evidence from the past, the present and projects probable futures. The plausibility of the scenarios created will depend on the quality of the evidence collected.

There are several methodological proposals for prospective scenarios, depending on the point of view of each author; however, for the present research, the six-pillar approach proposed by Sohail Inayatullah [6] is used, as its methodological process allows for the proper substantiation, creation and evaluation of prospective scenarios.

1.3 Objective

1.3.1 General

To project the morphological evolution of single-family dwellings towards the year 2100, using the scenario methodology for prospective studies, to obtain formal design strategies that can be adapted to a probable future.

1.3.2 Specifics

- Identify the case studies to be included in the proposed analytical process, through bibliographic review and audio-visual sources.
- Describe the morphology of the case studies through architectural surveys (floor plans, elevations, cuts, construction details, etc.) to establish the prospective design patterns and trends.
- Apply the scenario methodology for prospective studies, which makes it possible to project the morphological evolution of single-family dwellings towards the year 2100.

2. Methodology

> *"Methodology is the link between the subject and the object of knowledge. Without it, it is practically impossible to achieve the path that leads to scientific knowledge."*
>
> Anonymous [13]

The methodological process adopted for this research is based on the six pillars or stages for the study of Sohail Inayatullah's future [6], the scope of the research is documentary and descriptive.

2.1 Foresight scenarios: Six-pillar approach

This methodology stems from the foresight process proposed by Voros [14] and the prospective approach derived from Dator's Manoa school [15]. The six-pillars approach allows for the creation of scenarios based on scientific evidence from the past and present, using tools and methods articulated through praxis. The pillars are Planning (step 1), Anticipation (step 2), Timing the future (step 3), Deepening the future (step 4), Creating alternatives or scenarios (step 5), Transforming the future (step 6).

2.1.1 Step 1

To sketch the probable futures, the technique used in this stage is the futures triangle, through bibliographic and audio-visual documentary analysis, a probable future is established and described in the timeline with retrospective and prospective data, selecting more than 60 dwellings located in probable futures as case studies for the research.

2.1.2 Step 2

This stage has Molitor's emerging themes analysis as its main method [16]. The aim of the emerging themes analysis in the present research was to identify the main historical events that triggered the innovations or produced a historical disruption that changed the future of an event, thus allowing the identification of the branches of probable futures.

2.1.3 Step 3

In the temporisation of the future, the macro-patterns, meso-patterns and micro-patterns of change are studied to establish the timeline and its projection into the future. In this stage, the formal and constructive patterns were studied through the analysis of the façades of the selected case studies.

2.1.4 Step 4

The fourth pillar is to lay the logical foundations of the future, using stratified causal analysis or CLA [17]. This technique analyses the social vision of the future at different levels of complexity by exploring and verifying possible causes and refining the hypotheses of the scenarios to be proposed in the next stage.

2.1.5 Step 5

Because of the previous stages, the most relevant prospective scenarios are obtained, which are presented to a panel of experts so that the information collected and processed can be validated by consensus, thus obtaining the probable futures.

2.1.6 Step 6

In this last stage, the prospective scenarios are corrected and validated again with the experts through the 1) prognostic; 2) retrospective and 3) transcendental methods to resolve conflicts that may arise between the proposed scenarios.

3. Results

"One must know the past to understand the present and imagine the future"
Luis Sepúlveda [18]

The results obtained by applying the scenario methodology for prospective studies using the six-pillar approach are summarised below.

3.1 Synthesis of the past, analysis of the present, projection of the future

Proposing a probable future is very risky if the necessary evidence is not available; therefore, the theories of the future put forward by the researchers Jim Dator [19] Nikolai Kardashev [20] and Sohail Inayatullah [6] are considered as a starting hypothesis in the present research.

Jim Dator in his studies of the future proposes a foresight model (see **Figure 1**) in which it is observed that social change can be classified into four recurring groups of images, stories or policies about the effects of social change, and the trend indicates the following:

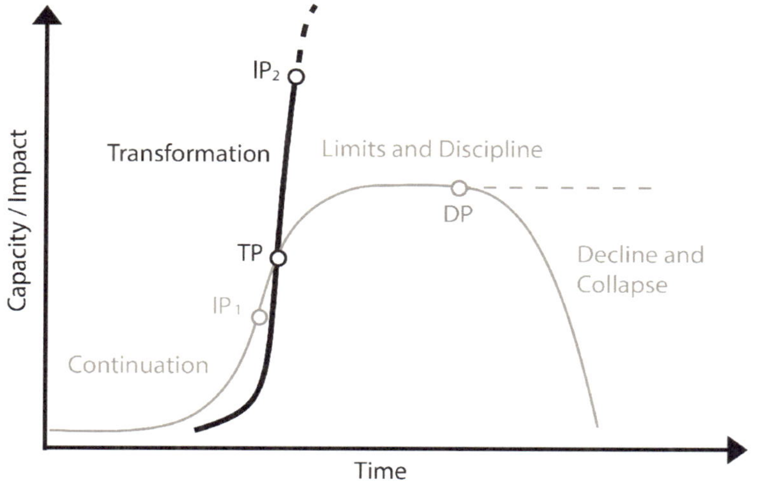

Figure 1.
The four Dator futures, growth-based alternative scenarios (GBAS).

1. Continuation (business as usual, further growth of the status quo).

2. Limits and discipline (behaviours to adapt to increasing internal or environmental limits).

3. Decay and collapse (system degradation or failure modes as crisis emerges).

4. Transformation (new technology, business or social factors that change the game).

Nikolai Kardashev proposes three types of civilisations on a scale that can be quantified in units of power (watts) and represented on an increasing logarithmic scale. **Figure 2** illustrates the types of civilisations per the proposed scale.

Sohail Inayatullah mentions that at the macro-global level, there are many images of the future; however, there are five archetypal ones;

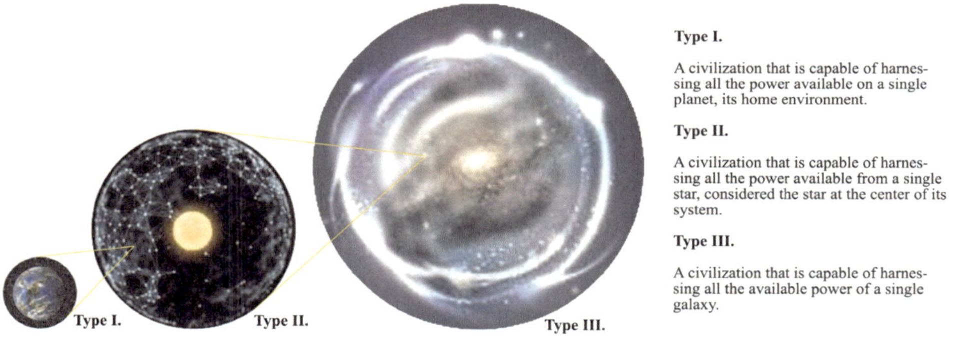

Figure 2.
Type of civilisations per the Nikolai Kardashev scale.

1. Evolution and progress.

2. Collapse.

3. Gaia.

4. Globalisation.

5. Back to the future.

Organising the above scenarios around the architectural field, four hypothetical futures are proposed, through which the present morphological study revolves, as shown in **Table 1**:

Hypothetical futures	**Vision of Sohail Inayatullah**	**Vision of Jim Dator**	**Vision of Nikolai Kardashev**	**Description of the future according to different visions**
Chaos	Collapse	Decline and Collapse		There is the idea that man has reached his limits; in fact, he has surpassed them: global inequality, fundamentalism, tribalism, nuclear holocaust, climatic disasters: all point to a worsening future.
	Back to the future			We need to go back to a simpler time, when the hierarchy was clearer, when technology was less disruptive, when the rules of the hierarchy were clear. The change is overwhelming: we have asked for direction and we need to get it back.
				System degradation or failure modes as crisis emerges.
Gaia	Gaia	Limits and Discipline	Civilisation TYPE III	The world is a garden and cultures are its flowers, we need social technologies to repair the damage we have done to ourselves, to each other and to nature, and above all to be ever more inclusive. The next leap in evolution is based on collaboration between men and women, humans and nature, and humans and technology.
				Behaviours to adapt to increasing internal or environmental limits.
				A civilisation that is capable of harnessing all the available power of a single galaxy.
Progress	Evolution and progress Globalisation	Continued	Civilisation TYPE I	More technology, man as the centre of the world and the belief in rationality.
				Barriers between nations and cultures can be removed if we achieve a free market system. Technology and the free movement of capital can make us all rich. Traditionalisms and dogmas are the barriers that prevent us from achieving a new world.

Hypothetical futures	Vision of Sohail Inayatullah	Vision of Jim Dator	Vision of Nikolai Kardashev	Description of the future according to different visions
				Business as usual, more growth of the status quo
				A civilisation that is able to harness all the power available on a single planet.
Evolution	Evolution and progress	Transformation	Civilisation TYPE II	More technology, man as the centre of the world and the belief in rationality.
				New technology, business or social game-changing factors.
				A civilisation that is able to harness all the power available from a single star.

Table 1.
Hypothetical Futures described from the visions of Sohail Inayatullah, Jim Datos and Nikolai Kardashev.

Once the hypothetical futures have been established, the methodology of the six pillars described above is applied, in each of which the evidence validating each proposed scenario is examined in depth, the results obtained in each step are set out below:

3.1.1 Planning

In **Figures 3–6**, using the futures triangle tool, it can be seen how actions from the past emerge are linked to milestones in the present and future projections are generated. Both actions and milestones respond to bibliographical evidence researched from secondary sources, while future projection uses audio-visual information such as series, documentaries or films that allow a general visualisation of probable futures.

In this first stage, it is fundamental to lay the theoretical foundations with scientific evidence of past and present events analysed through the bibliographical review of more than one hundred secondary sources including scientific articles, press releases and books, using the technique of critical content analysis, which allowed us to articulate possible future scenarios created through audio-visual sources such as films, documentaries or series, analysing more than sixty case studies of houses or residences created in hypothetical futures. It can be observed that the triangles that present the most evidence are those of chaos and evolution, while GAIA and progress present less evidence, which allows us to establish a first tendency that will be defined in the following stages.

It is also important to note that in the triangles of futures it is observed that there are gaps in the line of the present, which psychologically indicates the level of uncertainty and concern of humanity in the face of events that cannot be predicted. In this sense, the future of progress is the most favourable and feasible for humans, while the chaotic future is the least desirable

3.1.2 Anticipation

Through the analysis of the emerging themes, a timeline of historical milestones or events in the past and present is established, which allows for the delineation and

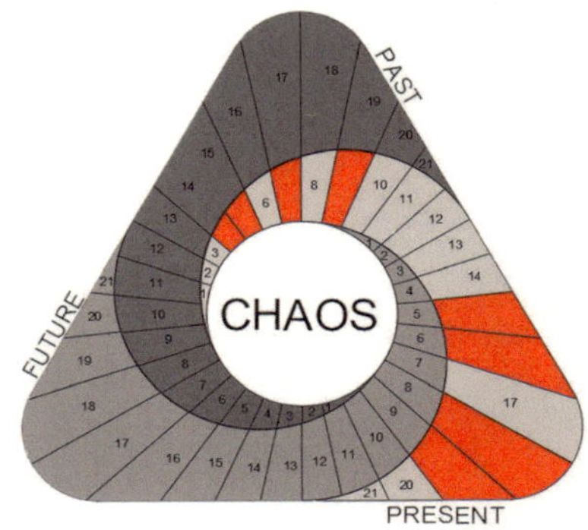

	actions of the **PAST**	milestones of **PRESENT**	projections of the **FUTURE**
1	World war II [21]	In the year 2020, North Korea has free access to military weapons, which would come to cause a war [22]	Serie: Defiance Film: After de Dark Video game: Metro Exodus
2	Developments of the first atomic bombs [22]	On February 27, 2022, the possibility of a nuclear conflict between Ukraine and NATO [23]	Video game: Metro 2033 Video game: Metro Exodus
3	Dropping of the atomic bomb on Hiroshima and Nagasaki, on August 6 and 9, 1945 respectively [22]	In June 2022, the war between Ukraine and Russia begins [24]	Film: Fall out Film: After de Dark Video game: Metro Exodus
4	Creation of the apocalyptic clock in 1947 by the NGO, within the bulletin of Scientists [25]		Film: Divergent Film: Insurgent Film: Divergent Film: Leal
5	Hydrogen bombs began their construction in 1951 [22]		Film: Fall out Film: After de Dark Video game: Metro Exodus
6	In the years 1951 to 1973 there was radioactive fallout that claimed the lives of between 4000 and 90000 people from cancer and skin burns [26]	The scientific community assures an imminent crisis due to the pollution caused by industries, including technology [27]	Film: Fall out Film: After de Dark
7	During the Cold War, in 1967, the Italian government built the first anti-atomic bunker on Monte Soratte [26]		Film: The Bunker Film: After de Dark
8	In 1970, it was determined that stress causes infertility due to hormonal imbalance [28]	Lockdown causes infertility [29].	Film: The Bunker
9	Chernobyl nuclear disaster, April 26, 1986, causing poisoning radioactive and warping [30]		Film: Fall out Film: After de Dark
10	In 1993, the Mount Soratte anti-atomic bunker was defined by the International Atomic Energy Agency as a valid example of how to intervene on an existing structure to ensure the continuity of a country in the event of atomic attacks [26]	In the year 2020 during COVID-19, the Survival Condo bunker was inhabited by millionaires [31].	Film: After de Dark
11	In the year 2000 it was proposed that the plant cell is totipotent atomic [32]	Creation of more durable transgenic plants [33]	Film: I am mother
12	In november 2021 the temperature reached 1.24°C above the average in the period from 1850 to 1900 [34]	In 2016 global warming reaches high levels and climate change is produced by greenhouse gases [35]	Film: The Island Film: The Wandering Earth
13	The progress of the internet began to be exponential in 1994 and therefore the resurgence of computing as it is known [36]	In the field of technological development, he states that robotics and artificial intelligence are changing our societies. These are fields considered to be the most economically developed in the 21st century. [37]	Film: Matrix Reloaded
14	Contamination due to non-compliance with regulations in the mining industry [38]	In 2016 global warming reaches high levels and climate change is produced by greenhouse gases [33]	Film: The Island Film: The Wandering Earth
15	With the industrial advances and wars that took place in the 20th century, the scarcity of resources began, forcing the US to create lists of imported products and create plans for their sustainability [34]		Film: Mad Max
16	By the year 1901, Hans Speman managed to divide the embryo of a salamander and make it develop, after that, in the year 1928, Speman determined that it is the nucleus of the cell that directs cell division [39]		Film: Blade Runner 2049 Film: I am mother
17	In 1997 the Dolly sheep was cloned, a similar event arose in 2002 when the first cloned cow was born in the Biosidus company in Argentina [40]	The first pig heart transplant to a man becomes a reality in the year 2022 [41]	Film: Blade Runner 2049 Film: I am mother
18	In 2004, Koreans cloned human embryos by transferring the nuclei of somatic cells taken from the ovary of donors to enucleated oocytes [35]		Film: Blade Runner 2049 Film: I am mother
19	In the year 1550 the little ice age occurred in the northern hemisphere, the first little ice age occurred in the year 1650, the second little ice age occurred in the year 1770 and finally the third little ice age in 1850 [42]		Film: Water world Film: The Wandering Earth

Figure 3.
Triangle of the future—CHAOS [21–42].

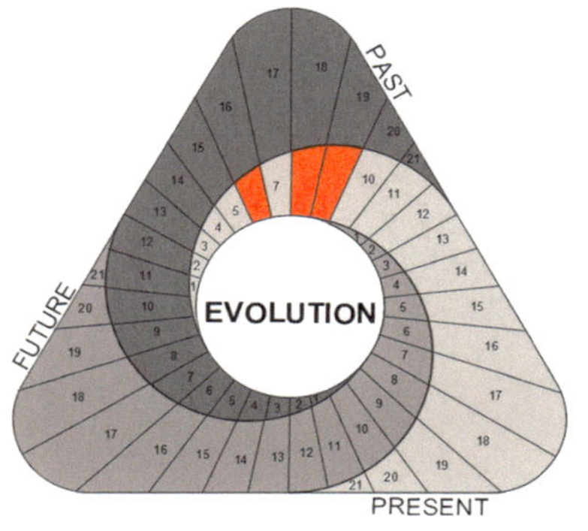

	actions of the **PAST**	milestones of **PRESENT**	projections of **FUTURE**
1	Satellite delivery Sputnik I 1957 [43].	Curiosity for Mars 2012[44].	Video game: Halo Reach Video game: Halo
2	Space travel Yuri Gagarin 1961 and Apolo 11 1969 [45].	Exploration of the solar system and creation of space stations [46].	Film: Thor 3- Ragnarock Film: Infinite Space Film: The wandering land
3	First characters to step on the moon Neil Armstrong and Buzz Aldrin [47].	Tourist space travel [48].	Video game: Halo Reach Video game: Halo Film: Prometheus
4	Self-propelled vehicles Lunokhod 1 y 2 [49].	Current industrial revolution [50].	Video game: Anno Film: Space Sweepers
5	Robotic geologists [51].	Search for microbial life with perseverance rover [52].	Film: Tron Film: Blade runner Film: Black Panther
6	Mission Apolo-Soyuz 1975 [53].		Film: Lost in Space
7	Sending humans to Mars [54].	Arrival of man on Mars [55].	Film: A space between us Film: Lost in Space
8	World war II [56].		Film: Dune
9	Internet creation [57].		Film: Oblivión : The time of oblivion
10	Spaceship delivery attempts [58].	Discovery of other possible habitable planets [59].	Video game: Halo
11	World war I [60].	War between Ukraine and Russia [61].	Video game: Halo Reach
12	Cold war [62].	Attacks in Paris 2015 [63].	Film: Star Wars- Star Wars
13	Bombing in Hiroshima and Nagasaki 1945 [64].	Environmental Pollution [65].	Film: After Earth Film: The future without oxygen Film: The perfect city
14	Attack of the world trade center 2001 [66].	Explosion in Beirut, Lebanon. [67].	Serie: The Expanse
15	The economic crisis of 1929 [68].	Crisis in Venezuela and migration of citizens 2018 [69].	Film: Elysium
16	Missile crisis, La Habana 1962 [70].	Bombing in Syria by Russia 2015 [71].	Serie: The 100
17	Holocaust in Alemania of 1933-1945 [72].	7.5 earthquake in Oaxaca 2020 [73].	Film: Avengers Infinity War
18	Rwandan genocide 1994 [74].	Overpopulation [75].	Film: Demolition Man
19	The Vietnam War 1955-1975 [76].	Attacks in Manchester and Barcelona 2017 [77].	Film: Ender's Game

Figure 4.
Triangle of the future—Evolution [43–77].

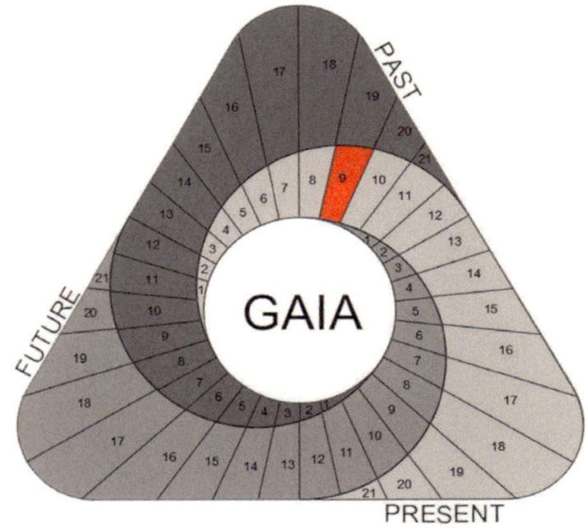

	actions of **PAST**	milestones of **PRESENT**	projections of **FUTURE**
1	A nuclear war would affect humans [78]	Has the human being begun to involute? [79]	Film: Divergent
2	Doomsday clock [80]	Limiting global warming to 1.5°C [81]	Documentary: Dream the future
3	Between 1880 and 2012, the global average temperature increased by 0.85 degrees Celsius [82]	Ecosystem change [83]	Film: Valerian and the city of a thousand planets
4	The extreme heat waves of 2003 and 2010 [84]	Heat waves cause serious health problems and high mortality rates [85]	Film: Star Wars Film: The Lord of the rings Film: Eragón
5	The oceans have warmed, the amount of snow and ice has decreased, and sea levels have risen [86]	In the year 2021 alone, we witnessed more than 50 serious floods in the world [87]	Film: Avatar II
6	The Industrial Revolution brought with it an increase in the gases responsible for the greenhouse effect. [88]	Famines and food shortages due to the Greenhouse Effect [89]	Film: Harry Potter
7	Nuclear weapons have only been used twice in war, in the 1945 bombings of Hiroshima and Nagasaki. [90]	In addition to cancer, nuclear bomb survivors suffer from heart disease, stroke or post-traumatic stress [91]	Film: The origin of the guardians
8	The ozone hole in Antarctica, discovered in the 70s [92]	The Montreal Protocol on Substances that Deplete the Ozone Layer [93]	Serie: Game of Thrones
9	Global emissions of carbon dioxide (CO2) have increased by almost 50% since 1990 [86]		Film: Black Panther
10	Overexploitation of the natural environment, cause of the loss of biodiversity [94]	The IPBES report mentions the Aichi Targets as transformative actions to protect biodiversity [95]	Film: Avatar

Figure 5.
Triangle of the future—GAIA [78–95].

correlation of actions taken by humankind over time, projecting hypothetical futures. **Figure 7** shows how events are intertwined, generating alternative prospective scenarios.

3.1.3 Temporisation of the future

Figure 8 shows the evolution of morphological patterns and the use of materials throughout history. Morphology is directly related to the technology and building materials available at the time.

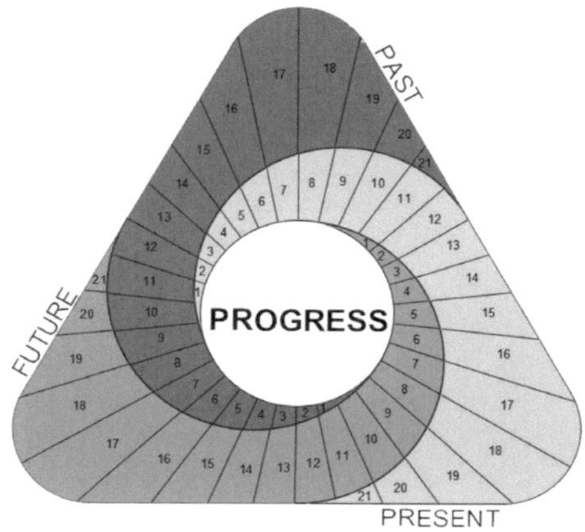

	actions of the **PAST**	milestones of **PRESENT**	projections of **FUTURE**
1	World War I [96]	Initial conflict between Russia and Ukraine [97]	Film: Coma
2	First contact with the moon [98]	James Space Telescope [99]	Documentary: The Architect
3	Creation of the Computer [100]	Quantum computing [101]	Film: The Sleeper
4	Discovery of Gravitational Waves [102]	The sound of a black hole [103]	Film: Body Double
5	World War I [104]	War between Russia and Ukraine [61]	Film: Men who did not love women
6	Great printing revolution [105]	Digital trust [106]	Film: Elysium
7	Technological advance in neurorehabilitation [107]	It is evaluated to create an innovative home automation prototype [108]	Film: The Avengers
8	Solar panels [109]	Internet 5G and 6G [110]	Film: RoboCop
9	Print [111]	3D printing with elastomeric materials [112]	Video game: Cyberpunk
10	Graphene [113]	COVID-19 [114]	Video game: Fallout
11	Industrial Revolution [115]	Attacks in Manchester and Barcelona [116]	Serie: 3%

Figure 6.
Triangle of the future—Progress [96–116].

Another important fact to highlight is the monumentality of the designs directly related to the gods at the beginning of the cultures and progressively public or residential buildings represent the architectural works of value, to carry out the analysis of all the buildings created and to be created would create several lines of research so this study focuses only on residential works. In this process of evolution of morphology, the first cultures used basic geometric shapes such as the triangle, the square and the trapezoid, and then incorporated the circle and arches, always guided by symmetry as a compositional principle. With the First Industrial Revolution, the creation of new materials and technological advances, man has been modifying the form since the contemporary age, the order and balance of symmetry changes for the complexity of

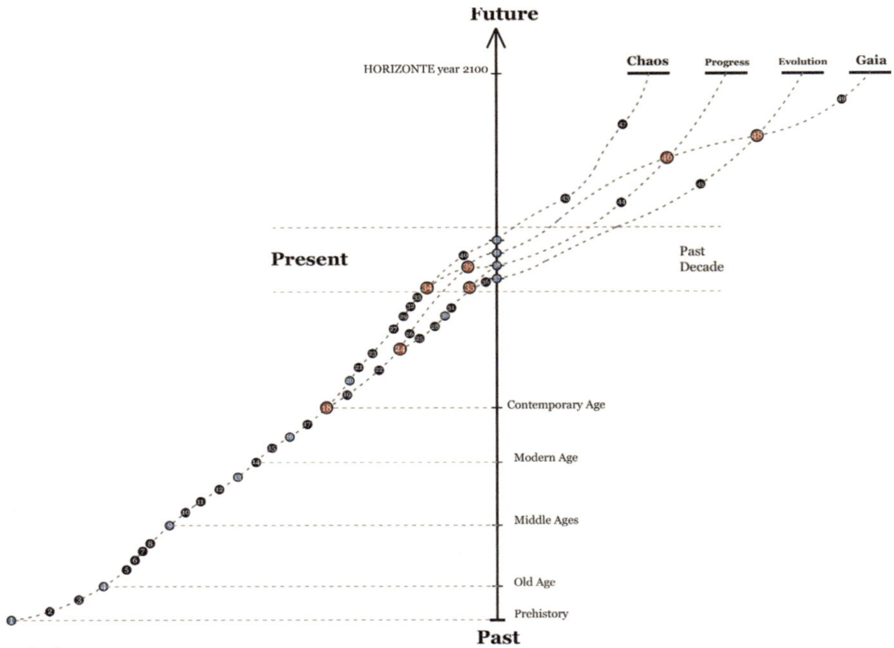

Symbology

1.- Stone carving - Stonehenge. 2.- Metals processing. 3.- Invention of the Wheel. 4.- **Invention of Scripture.** 5.- Mesopotamian Empire - Ziggurats - ramps.
6.- Egyptian Empire - Pyramids - lintel. 7.- Greek Empire - Temples - Columns - Poules. 8.- Roman Empire - Temples - Roman arch. 9.- **Appearance of Christ.**
10.- Romanesque style - barrel domes. 11.- Gothic style - ribbed domes - pointed arches. 12.- Castles - crusades - fortifications - plagues.
13.- **Renaissance** - man at the centre of the world. 14.- Discovery and colonisation of America . 15.- Styles Baroque - Rococo. 16.- **Rationalism.**
17.- Neoclassical style. 18.- **1st Industrial Revolution** - steam engines. 19.- French Revolution. 20.- **2nd Industrial Revolution** - electric power.
21.- World War I. 22.- Art Noveau - Modern Architecture. 23.- World War II - nuclear bomb. 24.- **III Industrial Revolution** - electronics and IT.
25.- The first trip to the moon. 26.- Organic, deconstructivist architecture, high tech, contemporary. 27.- Hole in the Ozone Layer. 28.- The message from Arecibo.
29.- Sustainable Development 30.- **Globalisation** 31.- Hubble Telescope Launch 32.- La clonación de la Oveja Dolly 33.- Twin Towers attack.
34.- **Global Warming.** 35.- **4th Industrial Revolution** - Artificial Intelligence. 36.- Mission persevere on Mars. 37.- Higgs boson discovery quantum computer construction. 38.- The development of nanotechnology. 39.- **The ODS.** 40.- The COVID-19. 41.- Weather Clock. 42.- War between Ukraine and Russia.
43.- Political tensions, the economic crisis, terrorist attacks in Europe and the USA have been the triggers for the start of World War III.
44.- The technological development of large companies such as TESLA, investment in research by the public and private sector have enabled sustainable progress.
45.- The colonisation of the Moon and Mars has prompted man to seek to conquer interstellar space in search of new sources of renewable energy.
46.- Following the evaluation of the ODS in 2030, the Agency for the **Sustainable Development of Humanity** was created in 2040 at the initiative of the ONU.
47.- After World War III with the use of atomic and chemical bombs, famine, pestilence and water shortages followed, where more than 3/4 of the world's population died.
48.- There is the **first intergalactic contact** whose message drives technological development and nature protection. 49.- Bio-construction on Earth begins.

Figure 7.
Chronology of historical events and projection of future scenarios.

parametric and organic designs, with 3D printing and artificial intelligence in the short- and medium-term architectural forms, will be totally unpredictable and complex, where the theory of chaos would gain value [117], because as has been observed throughout history we have gone from order to chaos and in the future, thanks to technological advances, from chaos to order.

3.1.4 Deepening the future

In this stage, the most plausible futures that could occur up to the year 2100 are defined and projected through stratified causal analysis. **Table 2** shows the different hypotheses proposed and levels of causality in the futures analysed, providing a more concrete vision of each one of them, which allows the information processed in the previous stages to be articulated and refined.

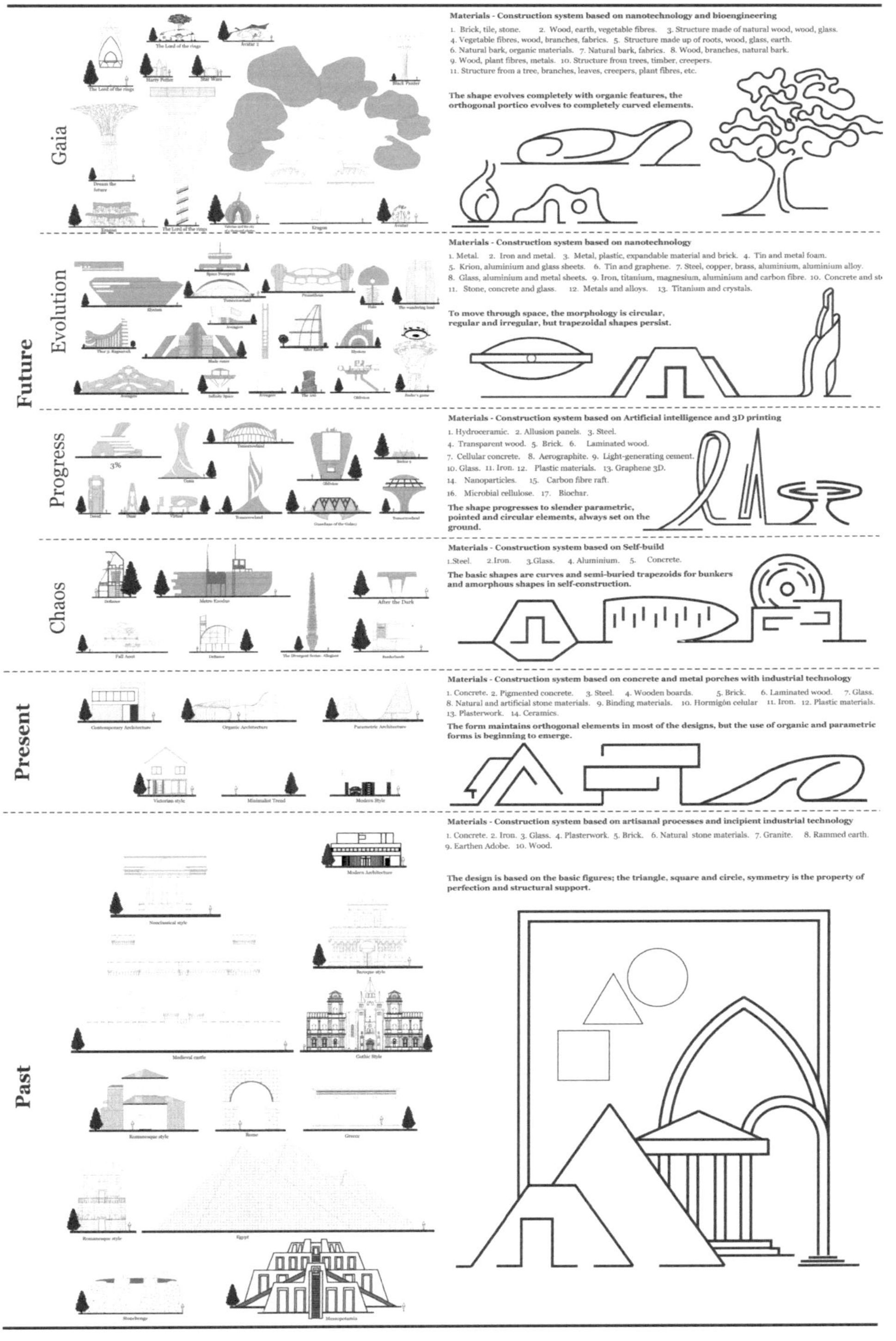

Figure 8.
Historical analysis, evolution and prospective of architectural morphology.

Level of CLA	Chaos	Progress	Evolution	Gaia
Litany	P: The need for protection in a chaotic world.	P: Building materials and systems are not sustainable.	P: Resources are about to run out on Earth.	P: Global warming will destroy the planet.
	S: Self-build housing.	S: Improve construction systems and materials.	S: New off-world energy sources must be sought.	S: Raise humanity's awareness of respect for the planet.
Systemic causes	P: The supply and labour chain is broken.	P: There is no technology that improves building systems.	P: Countries seeking to improve technology depleted renewable resources.	P: The process of economic and technological development of countries is not sustainable.
	S: Recycling materials and self-building	S: Generate research projects focused on sustainability.	S: Using technology to conquer interstellar space.	S: Promote bioclimatic and sustainable architectural projects.
Vision of the world	P: The world is chaotic and dangerous.	P: The technology is developing, but progress is slow.	P: The colonisation of other planets is a long way off.	P: If humanity does not stop polluting the planet it will cause irreversible damage.
	S: We have to build strength with what is available.	S: Develop technologically sustainable building materials and systems.	S: Develop technology for interstellar travel.	S: Promoting the sustainable development of the world's cities.
Myth/ metaphor	P: *'The danger is constant and it's out there'.*	P: *'Technology consumes a lot of non-renewable resources'.*	P: *'There is no life on other planets'.*	P: *'Climate change affects the natural process of life on Earth'.*
	S: *'Strengthening housing strengthens survival'.*	S: *'Technology development must be sustainable'.*	S: *'Extraterrestrial technology would exponentially enhance our own'.*	S: *'Generating a symbiosis between architecture, energy and nature is possible'.*
Conclusion	Form is not important in the face of the need for protection.	Form is based on function and purity of materials.	The shape and materials must allow for life outside the Earth.	Morphology is governed by the form and laws of nature.

Table 2.
Stratified causal analysis of proposed scenarios.

3.1.5 Creating alternatives or scenarios

In this stage, morphological proposals of the architectural works analysed are exhibited through different audio-visual sources that could occur in the proposed future scenarios.

3.1.6 Chaos

In this scenario, the tendency is towards an apocalyptic future, the product of nuclear, chemical and biological wars between world powers, motivated by the domination and extraction of natural resources. Faced with this scenario, dwellings are transformed into shelters whose morphology responds to the need for protection and the availability of resources at the time, with the semi-buried bunker and self-construction with recycled materials predominating, the morphological basis of design can be abstracted from the massiveness of the ziggurats, pyramids or mastabas with semi-buried chambers. The predominant materials are: reinforced concrete, iron, aluminium, glass, plastic and recycled wood. **Figure 9** shows proposals for dwellings whose morphology responds to this future.

3.1.7 Progress

In the scenario of progress, humanity has reached Type I civilisation, in which with technological progress mankind takes advantage of all the planet's resources in a sustainable way, houses are built with synthetic materials produced with high technology, nanomaterials, as well as artificial intelligence, play a primordial role in this future, morphologically the houses have parametric, and slender features with large glazed surfaces as can be seen in **Figure 10**.

Figure 9.
Examples of residences in a future of chaos.

Figure 10.
Examples of residences in a future of progress.

3.1.8 Evolution

With the horizon of the year 2100, the scenario of evolution is located in the Type II civilisation where humanity has managed to conquer and take advantage of the resources of the solar system, space travel and the colonisation of planets have taken place thanks to the technological advances of the time, the morphology of the dwellings responds to the need to live in interstellar space, the materials are highly resistant, nanotechnology, as in the previous scenario, plays a transcendental role in construction, the shapes are aerodynamic curves, the orthogonal portico is limited, and tubular or dome-type structures are the trend in this scenario. **Figure 11** shows alternatives for housing in this future.

3.1.9 Gaia

The symbiosis between nature, man and architecture is the trend of this scenario, man has learned from past experiences and has managed to understand that nature always wins in any battle, the respect between communities and the environment is unique, the resources and materials are organic and sustainable, the morphology of housing is based on completely organic forms, whose design base could be the current organic and parametric architecture but ancestral construction systems. **Figure 12** shows some proposals for housing whose morphology responds to this future.

Figure 11.
Examples of residences in a future of evolution.

Figure 12.
Examples of future GAIA residences.

3.1.10 Transforming the future

The prognostic technique, hindsight and the transcendence method allow the proposed futures to be validated through experts. According to Escobar Jasmine, expert judgement is defined as an informed opinion of people with a track record in the subject matter, who are recognised by others as qualified experts in the subject matter, and who can provide information, evidence, judgements and assessments. The number of judges to be used in a trial depends on the level of expertise and diversity of knowledge; however, the decision on what number of experts is appropriate varies among authors, with some suggesting a range of two to twenty experts, while others state that ten would provide a reliable estimate of the validity of the information [118]. The selection of the number and quality of experts were made as follows:

1. An expert profile was determined: Architect or researcher with a Master's or PhD degree, who has a minimum of five consecutive years of experience in research on morphological design of single-family dwellings proven through publications.

2. From a pool of more than 50 potential candidates from all over the world, who were contacted by email requesting their participation, 30 candidates responded affirmatively with their willingness to participate.

3. The 30 experts were sent an interview questionnaire with key questions to validate the proposed scenarios, 19 completed the questionnaire and only 13 completed it without any errors. Therefore, taking the recommendation of Escobar Jasmine, the information collected was considered valid.

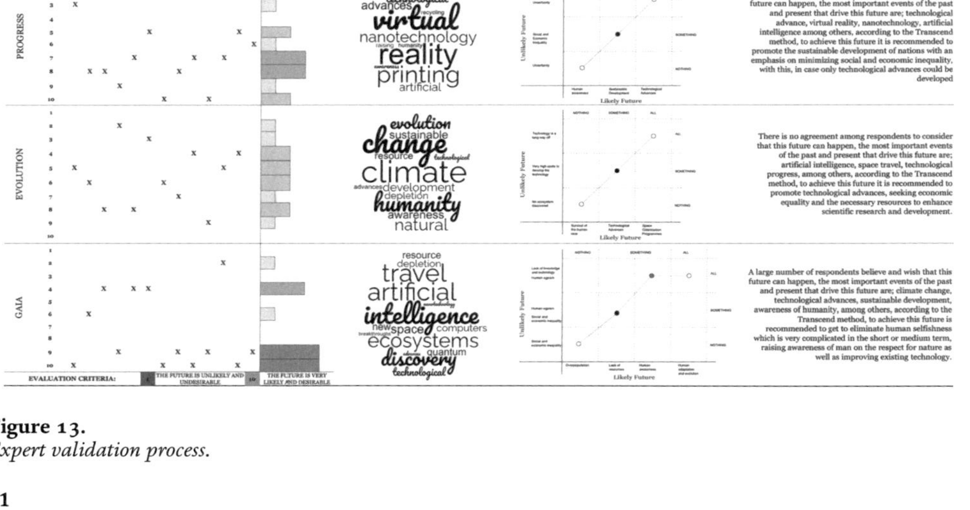

Figure 13.
Expert validation process.

Figure 13 shows this validation, where chaos and GAIA are the most extreme undesirable and desirable futures, respectively, while progress and evolution are the average and possibly achievable futures by their events and forecasts.

4. Conclusions

By way of conclusion to the study, the following considerations are made:

1. The study of the future is at an incipient stage, and the uncertainty of mankind in the face of uncertainty is still great, however, under the position of Jules Verne; what one man can imagine, another man can make real, and it is considered that the present analysis will give a glimpse of probable futures.

2. The methodology of the study of the future under the approach of the six pillars allows at each stage to deepen the analysis and synthesize to finally validate it through experts, managing to consolidate and create a research with adequate methodological and scientific basis.

3. Because of the research, **Figure 14** shows the morphological evolution, it is considered that by the year 2100 the form will evolve to curved parameters, complying with the laws that govern nature, this is supported by the theory of John Archibald [119] who states that "Space-time tells matter how to move; matter tells space-time how to curve" and in this case architecture is the art of creating or modelling space to satisfy the needs of man.

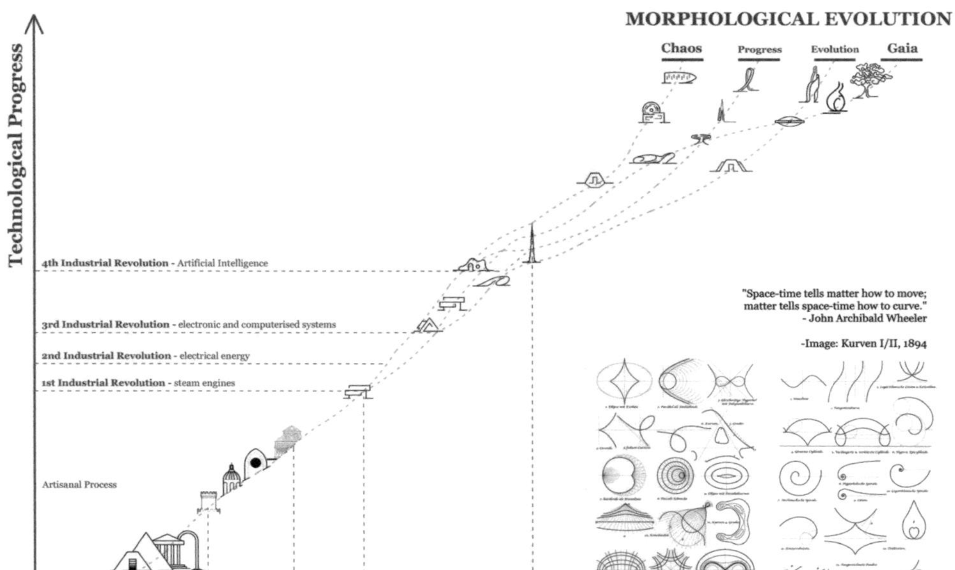

Figure 14.
Evolution analysis and morphological projective of residences in the year 2100.

Acknowledgements

This article is part of the Doctoral Thesis of the Doctoral Programme in Architecture of the National University of Cordoba, articulated to the formative research of the Catholic University of Cuenca, for this we thank the authorities and students of the third cycles who supported the process, especially Santiago Diaz, Victor Gomezcoello, Melany Guaman and Kevin Nieves.

Author details

Marco Avila-Calle[1,2]*, Jorge Toledo-Toledo[1] and Federico Córdova-González[1]

1 Catholic University of Cuenca, Cuenca, Ecuador

2 National University of Córdoba, Córdoba, Ecuador

*Address all correspondence to: mavila@ucacue.edu.ec

References

[1] Todo lo que una persona puede imaginar, otras podrán hacerlo realidad. Julio Verne | Biblioteca de Torredelcampo [Internet]. [cited 2022 Aug 22]. Available from: https://bibliotecadetorredelcampo.wordpress.com/2016/02/08/todo-lo-que-una-persona-puede-imaginar-otras-podran-hacerlo-realidad-julio-verne/

[2] El Triángulo de Futuros - Strategic Foresight [Internet]. [cited 2022 Aug 22]. Available from: https://www.strategicforesight.es/blog/el-triangulo-de-futuros/

[3] Griffiths M, Levy N, Timpe K. The Routledge Companion to Free Will. New York: Routledge; 2016. pp. 364-373

[4] Benigno M, Calle A, Santiago M, Quezada O, Humberto D, Torres C. La arquitectura adaptativa, una visión integral de la obra arquitectónica sustentable. 35-54

[5] Yankovskaya YS, Merenkov AV. Image and morphology in modern theory of architecture. IOP Conference Series: Material Science Engineering. 2017;**262**:012134

[6] Inayatullah S. Estudios del futuro : teorías y metodologías. Hay Futur Visiones para un mundo Mejor [Internet]. 2012;39–71. Available from: https://www.bbvaopenmind.com/wp-content/uploads/2013/10/Hay-futuro_visiones-para-un-mundo-mejor_BBVA.pdf

[7] Interdisciplinar D, Interpretación YSU. Proyectual del Estudiante complexity in architectural morphogenesis ; Interdisciplinary dialogue and its interpretation as a method in the student ' s projectual process Juan Andrés Sánchez García. 2019; 85-99

[8] Diseño, esteticidad y discurso - Libros de Universidades [Internet]. [cited 2022 Aug 22]. Available from: https://ulibros.com/diseno-esteticidad-y-discurso-tcdm4.html

[9] Geertz C. La interpretación de las culturas. México: Editorial GEDISA; 1987. p. 387

[10] Aguirre Collahuazo JP. LECTURE HALL III of the University of Alicante (1998–2000), Architect Javier García-Solera Vera Architectural project analysis. Estoa. 2016;**005**(008):41-56

[11] Calle MA, López GC. Ecuador: Perspectives of the past, present and future: A multi-criteria approach to social evolution. Ecuador: Perspectives of the Past, Present and Future: A Multi-Criteria Approach to Social Evolution. 2021;1-257

[12] Avc B. Metodología de los escenarios para estudios prospectivos. Ing e Investig. 2011;**44**:26-35

[13] Diferencia entre Método y Metodología - Metodología de la Investigación [Internet]. [cited 2022 Aug 22]. Available from: https://sites.google.com/site/51300008metodologia/diferencia-entre-metodo-y-metodologia

[14] Voros J. A generic foresight process framework. Foresight. 2003;**5**(3):10-21

[15] Dator J. Introduction: The future lies behind! thirty years of teaching futures studies. American Behavioral Science. 1998;**42**(3):298-319

[16] Molitor GTT. Molitor forecasting model: Key dimensions for plotting the 'Patterns of Change'. Journal of Future Study. 2003;**8**(1):61-72

[17] Kidd W, Reynolds S. Introduction: To the reader. Contemporary French Cultural Studies. 2014;**2014**:3-9

[18] Santisteban A, Pagès J. Aprender el pasado, comprender el presente y construir el futuro: lecciones de una investigación sobre la historia enseñada. XII Jornadas Interes Hist Dep Hist Fac Humanidades y Cent Reg Univ Bariloche Univ Nac del Comahue, San Carlos Bariloche, 2009;23. Available from: https://www.aacademica.org/000-008/843

[19] Los cuatro futuros de Dator: la guía prospectiva [Internet]. [cited 2022 Aug 22]. Available from: https://www.foresightguide.com/dator-four-futures/

[20] Escala de Kardashov (o qué es civilización I, II, III, IV y V) - Epicurea [Internet]. [cited 2022 Sep 5]. Available from: https://epicurea.org/escala-de-kardashov-o-que-es-civilizacion-i-ii-iii-iv-y

[21] Hernández J. Proyecciones y resultados: El trotskismo norteamericano y las consecuencias de la Segunda Guerra Mundial (1943-1947). Anuario de la Escuela de Historia Virtual. 2016;**9**:143-147

[22] Sola C, Sotelo M. El difícil camino hacia el control de la energía nuclear. Sustain. 2020;**4**(1):1-9

[23] Padilla LA. Neutralidad y Equilibrio de Poder en el Conflicto de Ucrania. Revista Costarricense de Política Exterior. 2004;**1**(22):33-52

[24] Cardone IJ. El Conflicto en Ucrania: los Intereses de las Grandes Potencias y los Perdedores de Siempre. Conjunt Glob. 2014;**3**(3):140-148

[25] Quispe D. Iusnaturalismo post positivista y propuesta de ley universal de extinción de dominio de bienes relacionados con actividades en armas nucleares y de destrucción masiva. 23-46

[26] Zamrodah Y. Programa internacional sobre los efectos del accidente de chernobyl en la. Salud. 2016;**15**(2):1-23

[27] Gonzalez Martinez A. Efectos contaminantes de industrias agroaumentarias: Mataderos E industrias cárnicas. Antonio González Martínez. 1993;**81**:81-92

[28] Veiga MC, Puccio MF, Tamburelli V. Qué relación existe entre estrés, ansiedad e infertilidad. Reproduccion. 2013;**28**(3):79-89

[29] Albeitawi S, Al-Alami Z, Khamaiseh K, Al Mehaisen L, Khamees A, Hamadneh J. Conception Preferences during COVID-19 Pandemic Lockdowns. Behavioural Science (Basel). 2022;**12**(5):4-11

[30] Sanchez C. Vista de El accidente de Chernóbil.pdf

[31] Garrett B. Doomsday preppers and the architecture of dread. Geoforum. 2021;**127**:401-411

[32] Villarreal LM. Totipotencia celular: Una revisión y aplicación del concepto. Botanical Sciences. 2014;**92**:318

[33] Carrasco JF. Plantas transgénicas. 1999;(3)

[34] Rodriguez D. Estudio termodinamico de la integración de un ciclo Kalina para aprovechamiento del calor residual en sistemas para generación termosolar [Tesis de pregrado]. España: Universidad Politécnica de Madrid; 2019. Available

from: https://oa.upm.es/57273/1/TFG_DIEGO_RODRIGUEZ_CUESTA.pdf

[35] Benavides H, León G. Gases de efecto invernadero related papers. Academia. 2007;**08**:35-47

[36] Saavedra FE, Vásquez JP. Las diferencias Norte-Sur en el debate ambiental global. Universum. 2017;**32**: 63-80

[37] Larios HM. Nuevo avance hacia la clonación terapéutica: se reaviva la polémica humana: se reaviva la polémica cloning: the controversy is revived. 2004

[38] Cardona Cardona FA, Políticos P en RI y E. Estudio de la minería ilegal aurífera: financiadora del crimen organizado especializado (Coe) en el occidente antioqueño 2013 – 2015. Caso Buriticá. Alcaldía Buriticá (2016) Presentación Buriticá Recuper http// www.buritica-antio quia.gov.co/prese ntacion.shtml. 2017;1-64

[39] Pettinari J. El camino hacia la clonación. Química Viva. 2004;**3**(1):2-5

[40] Núñez CM. El caso de la oveja 'Dolly'. Bioètica debat Trib abierta del Inst Borja Bioètica. 1997;**8**:8-9

[41] The xenotransplant patient who died received a heart infected with a pig virus | MIT Technology Review [Internet]. [cited 2022 Aug 23]. Available from: h ttps://www.technologyreview.com/ 2022/05/04/1051725/xenotransplant-pa tient-died-received-heart-infected-with-pig-virus/

[42] Garcia L. Cambios climáticos y efectos ambientales. 2008;56-66

[43] Cavallaro U. Triggers the USSR–USA Competition: A Story of Puffery vs. the Pragmatic. 2018;1-64

[44] Barceló M. El viaje a marte sin lanzaderas. 2012;1

[45] Patel. Historia de los trajes espaciales. 2019;9-25

[46] Calle A. La exploración del Sistema Solar. Rev Ciencias. 2013;**1**:21-28

[47] Duran X. Viajes literarios a la luna. 2016;**15**(2):1-23

[48] Barragán Acero V, Bonilla Arias MA, Gómez Arias J, Martínez Pérez AG, Moreno Millán KN, Navarrete Acuña V, et al. Análisis de los destinos turísticos del futuro. Punto de vista. 2019;**10**(15): 8-9

[49] Guarnizo Marin JG, Bautista Díaz D, Sierra Torres JS. Una revisión sobre la evolución de la robótica móvil. Una revisión sobre la Evol la robótica móvil. 2021;**2021**:1-8

[50] Orellana-Daube DF. El efecto global de la actual revolución tecnológica 4a revolución industrial y la industria 4.0 en acción. Rev GEON (Gestión, Organ y Negocios). 2020;**7**(2):1-24

[51] Astrofísica I, Andalucia D. ¡ DE 7 A 200 MIL: EL SISTEMA SOLAR CRECE ! René Duffard Introducción: Cometas, Trans-Neptunianos y satélites helados

[52] Clos A. Misiones a Marte. 2020. Available from: https://www.lanasa.net/misiones/marte

[53] Novgorod O. Cronología histórica de la Federación Rusa; 483-492

[54] Sirviente Alonso Carlos. Marte: construcción del primer asentamiento humano. 2017. Available from: https://oa.upm.es/47828/

[55] Martínez LV. CONFERENCIA The many paths toward Mars. 2017;487-498

[56] Beevor A. La Segunda Guerra Mundial. Colegio24hs [Internet]. 2004; 11. Available from: http://www.clasesh istoria.com/2guerramundial/movil/2gm-movil.pdf

[57] Castells M. Internet y la sociedad red. Conf Present del Programa Dr sobre la Soc la Inf y el Conoc [Internet]. 2001; 1-13. Available from: https://engage.intel.com/servlet/JiveServlet/downloadBody/26111-102-1-31790/INTERNET Y LA SOCIEDAD RED.pdf

[58] Enlae A, Ambrosio P, Ruiz JP, Cosmos ELU. Tema: Los viajes es aciales; 5-6

[59] Barra PN, Planetaria H. Busqueda de vida en otros planetas. 2018;46-55

[60] Ocaña J. La Primera Guerra Mundial, 1914-1918. Clío Hist Hist Teaching. 2001;**24**:5

[61] Perez V, Ramírez M, Tafur MA. ¿Es posible analizar la guerra entre Rusia y Ucrania desde el paradigma realista y liberal? 2022;(May). Available from: https://www.researchgate.net/publication/360311429

[62] McMahon J. et al. Introducción. Vol. 1, Robert J. McMahon La Guerra Fría Una Breve Introducción. 2003

[63] Roy J. Consecuencias de los ataques de parís. 2015

[64] González SL. Hiroshima y Nagasaki : memorias entre el conflicto , la reconciliación y los retos pacifistas. 2022; 191-203

[65] Bird L. La contaminación ambiental como factor determinante de la salud. Journal of Business Continental Emergency Plan. 2022;**15**(4): 300-301

[66] Antognazzi I. El atentado contra las Torres Gemelas NY 11/09/2001 visto a través de la teoría de la historia. 2005; 9-25

[67] Fern T, Vard R. Los estallidos del puerto de Beirut y sus efectos a un año de la catástrofe. 2:1-9

[68] Serrano SM. La crisis económica de 1929: Rosevelt y el new deal. Rev la Fac Ciencias Soc y Jurídicas Elche. 2010;**1**: 112-130

[69] Callata E, Elizabeth A, Ibarra M, Rodríguez C, Crasto TC, Álvarez MR. El éxodo Venezolano: Entre el exilio y la emigración. Migraciones. 2017;**4**:173

[70] Marín A. Crisis de los misiles de 1962: génesis de los cambios en las determinaciones en el Sistema Internacional. Perspect en Intel. 2016;**8**(17):21-32

[71] Ghotme R, Ripoll A. Las relaciones internacionales de la guerra civil Siria: Estados Unidos y Rusia en la lucha por el poder internacional. Rev Relac Int Estrateg y Segur. 2014;**9**(2): 77-102

[72] Ras M. Seminario de investigación : El Holocausto, la Solución Final y. 2016

[73] Ortíz Soto D, Reinoso Angulo E, Villalobos Ruíz JA. Daños observados en negocios por el sismo del 23 de junio de 2020 en Oaxaca. Acad Journals. 2020; **2020**:656-662

[74] Gonzalo SB, Lavín E, Rukebesha E. How the newspapers El País, Le Monde, Le Soir and the New York times treated information on the genocide in Rwanda in 1994. Palabra Clave. 2018;**21**(4):1214-1244

[75] Miranda ME. El Fenómeno Actual de Sobrepoblación Humana y los Retos que

Plantea la Situación Demográfica para el Perú. Derecho Soc. 2011;**37**:325-330

[76] Nievas F. La guerra de Vietnam (1964–1975). Contrainsurgencia. 2004;**6** (1):1-6

[77] Souza J. ATENTADOS 2017: SOMALIA, MANCHESTER Y BARCELONA. 26/27. 2015;I(Principio activo y prestación ortoprotésica):40

[78] Una guerra nuclear también tendría consecuencias catastróficas para el clima [Internet]. [cited 2022 Aug 17]. Available from: https://www.nationalgeographic.com.es/ciencia/guerra-nuclear-tambien-tendria-consecuencias-catastroficas-para-clima_17999

[79] ¿Ha empezado a involucionar el ser humano? [Internet]. [cited 2022 Aug 17]. Available from: https://pijamasurf.com/2018/06/ha_empezado_a_involucionar_el_ser_humano/

[80] Los científicos del 'reloj del fin del mundo' fijan sus manecillas a cien segundos de la catástrofe | Ciencia | EL PAÍS [Internet]. [cited 2022 Aug 17]. Available from: https://elpais.com/ciencia/2021-01-27/los-cientificos-del-reloj-del-fin-del-mundo-fijan-sus-manecillas-a-cien-segundos-de-la-catastrofe.html

[81] El cambio climático y sus implicaciones en la salud humana* Climate Change and Its Implications for Human Health Tania Angélica Cuadros Cagua**2. Available from: https://doi.org/10.11144/Javeriana

[82] Nations U. Cambio climático | Naciones Unidas. [cited 2022 Aug 17]; Available from: https://www.un.org/es/global-issues/climate-change

[83] cambio climático | Oxfam International [Internet]. [cited 2022 Aug 17]. Available from: https://www.oxfam.org/es/taxonomy/term/381

[84] Olas de calor y olas de greenwashing | Agora: Inteligencia Colectiva para la Sostenibilidad [Internet]. [cited 2022 Aug 17]. Available from: https://www.agorarsc.org/olas-de-calor-y-olas-de-greenwashing/

[85] Olas de calor y salud - OPS/OMS | Organización Panamericana de la Salud [Internet]. [cited 2022 Aug 17]. Available from: https://www.paho.org/es/campanas/olas-calor-salud

[86] Cambio climático – Desarrollo Sostenible [Internet]. [cited 2022 Aug 17]. Available from: https://www.un.org/sustainabledevelopment/es/climate-change-2/

[87] Inundaciones en el mundo costaron u$s82.000 millones en 2021 [Internet]. [cited 2022 Aug 17]. Available from: https://www.ambito.com/informacion-general/inundaciones/el-mundo-costaron-us82000-millones-2021-n5404873

[88] ¿Sabes cuándo se inició el cambio climático? | Palmitos Park [Internet]. [cited 2022 Aug 17]. Available from: https://www.palmitospark.es/blog/conservacion-y-sostenibilidad/inicio-de-calentamiento-global/

[89] ¿Cuáles son las consecuencias del efecto invernadero? | Ingredientes que Suman [Internet]. [cited 2022 Aug 17]. Available from: https://blog.oxfamintermon.org/cuales-son-las-consecuencias-del-efecto-invernadero/

[90] Armas Nucleares – UNODA [Internet]. [cited 2022 Aug 17]. Available from: https://www.un.org/disarmament/es/adm/nuclear-weapons/

[91] Hiroshima y Nagasaki, 70 años de efectos secundarios | Ciencia | EL PAÍS

[Internet]. [cited 2022 Aug 17]. Available from: https://elpais.com/elpais/2015/08/08/ciencia/1439021562

[92] Capa de ozono: por qué el agujero de la Antártida es el más grande de la última década (y no es necesariamente una mala señal) - BBC News Mundo [Internet]. [cited 2022 Aug 17]. Available from: https://www.bbc.com/mundo/noticias-54472884

[93] Nations U. Día Internacional de la Preservación de la Capa de Ozono ES | Naciones Unidas. [cited 2022 Aug 17]; Available from: https://www.un.org/es/observances/ozone-day

[94] Las causas de la pérdida de biodiversidad - Fundación Aquae [Internet]. [cited 2022 Aug 17]. Available from: https://www.fundacionaquae.org/wiki/causas-perdida-biodiversidad/

[95] Echeverry-Galvis MÁ, Unda M, Bravo MP, García N, Rubiano-Pinzón GA, Palomino JV. Plan de Acción en Biodiversidad de la PNGIBSE, Metas Aichi y los Objetivos de Desarrollo Sostenible en Colombia: análisis y concordancia en el discurso entre metas propuestas para 2020. Gestión y Ambiente. 2019;**22**(2):207-234

[96] Schmitt C. El orden del mundo después de la segunda guerra mundial. Rev Estud políticos. 1962;**122**:19-38

[97] Baqués J. El papel de Rusia en el conflicto de Ucrania: ¿La guerra híbrida de las grandes potencias? Rev Estud en Segur Int. 2015;**1**(1)

[98] Aghdasi F, Chao A, Lali M, Vasmate V. LunaRevolution-Role of the Moon in the Future of Human Space Activity LunaRevolution-Role of the Moon in the Future of Human Space Activity. 2016

[99] Garcia J. Grandes telescopios ópticos-Del catalejo al Telescopio Espacial James Webb [Internet]. ACTA representa en CEDRO los intereses de los autores científico-técnicos y académicos; 2022. Available from: https://www.acta.es/medios/articulos/ciencias_y_tecnologia/110001.pdf [Cited: 17 August 2022]

[100] Becker FG, Cleary M, Team RM, Holtermann H, The D, Agenda N, et al. Syria Stud. 2015;**7**(1):37-72

[101] Rosas Ortiz O. ¿ Computación Cuántica ? Tendencias actuales la Física en México [Internet]. 2007; Available from: http://www.fis.cinvestav.mx/~orosas/difusion/qcomp.pdf

[102] Compeán HG. Ondas gravitacionales. Investig Cienc. 2013;**IX**(31):24-38

[103] Pelat A, Gautier F, Conlon SC, Semperlotti F. The Acoustic Black Hole: a review of theory and applications. Journal of Sound Vibration. 2020;**2020**:115316

[104] Ortiz-De-Urbina P. La Primera Guerra Mundial y sus consecuencias: La imagen de Alemania en España a partir de 1914. Rev Filol Alem. 2007;**15**:193-206

[105] Goldman N. pública - y - debate - constitucional - en el - Río - de - la - Plata - (1810-1827). 2000;9–20

[106] Garzón G, Ruiz G, Juárez B. Análisis de la confianza , lealtad e intención de compra digital de los consumidores post-millennials. Rev Espac. 2020;**41**(34):141-154

[107] Gutiérrez-Martínez J, Núñez-Gaona MA, Carrillo-Mora P. Avances tecnológicos en neurorrehabilitación. Rev Investig Clin. 2014;**66**:8-23

[108] Rodriguez A, Fernández M. La casa inteligente. Univ CARLOS lll MADRIL. 2012;**8**:2-4

[109] Arencibia 2016. La importancia del uso de paneles solares en la generación de energía eléctrica Por Gustavo Arencibia-Carballo. Redvet. 2016;**17**(6):2

[110] Gustavsson U, Frenger P, Fager C, Eriksson T, Zirath H, Dielacher F, et al. Implementation challenges and opportunities in beyond-5G and 6G communication. IEEE Journal of Microwaves. 2021;**1**(1):86-100

[111] Spell LM, Medina JT. Historia de la imprenta en los antiguos dominios espanoles de America y Oceania. Hisp Am Hist Rev. 1960;**40**(1):139

[112] León M, Marcos-Fernández Á, Rodríguez-Hernández J. Impresión 3D con materiales elastoméricos. Rev plásticos Mod Cienc y Tecnol polímeros. 2019;**118**(747):2

[113] Medrano M, Guadalupe M, Rosu HC, Torres Gónzales LA. Disponible en. Acta Univ. 2012;**22**(3):20-23

[114] Pranggono B, Arabo A. COVID -19 pandemic cybersecurity issues. Internet Technology Letters. 2021;**4**(2):4-9

[115] Galbiatti M, Rural EM. Revolución Industrial Revolución Industrial. Ed Santillana. 2013;**10**(8):1-18

[116] Souza J. Trabajo de Fin de Grado. 26/27. 2015; I(Principio activo y prestación ortoprotésica):40

[117] Pidal M. La Teoría del Caos en las Organizaciones. Cuad Univ. 2009;**2009**: 29-33

[118] Escobar-p J. Validez de contenido y juicio de expertos: Una aproximación a su utilización. 2017

[119] Wheeler JA, Ford K, Rigden JS. Geons, black holes and quantum foam: A life in physics. American Journal of Physics. 2000;**68**(6):584-585

Section 3

Exploring the Potential of Green and Digital Regeneration

Chapter 6

Regenerate Corviale

Giovanni Caudo

Abstract

The Corviale is one of Rome's most problematic neighbourhoods. This neighbourhood's project to transform buildings is currently underway, sponsored by the Lazio Region. The 'Quarto Piano' was intended for services, community spaces and professional studios but has been converted into illegal apartments. The census had counted 135 households, but the project realised 103 apartments, a number lower than the number of households. The article reconstructs the activities carried out by the Roma Tre University research working group to complete the transformation of the building and focuses on the social support activities carried out with the residents. Numerous impacts, including intangible ones, come into play and characterise housing policy today; at least, this is the thesis we want to support in this article. The article focuses precisely on these impacts by highlighting that today's housing issue represents a spectrum of situations far beyond the housing. The issue of housing, considered in the light of this case study, where we can not only look at the tradition but also the innovation of housing policy, and described in the final chapter, highlights some lines of action that can have a general value that goes far beyond the case reproduced here.

Keywords: housing, public housing, urban regeneration, heritage, adaptive reuse

1. Introduction

Corviale is Rome's most famous public housing district and enjoys an international reputation. Its size has always caught the interest of architects and the imagination of citizens. Its main features are length, almost a kilometre, and the concentration of all accommodations in a single building. For these reasons, it has already received much attention from researchers in the field of housing studies. Since 2019, a vital regeneration intervention has been underway for the 'Quarto Piano' of the building. Over time, a hundred families have settled illegally, converting the spaces intended for public services, shops and offices into self-built flats. Years after the start of this self-build process, the Lazio region has financed an intervention programme to stabilise the change of use. The self-built houses are being replaced by new houses designed from scratch. It is an intervention that is only partly constructional. Numerous impacts, including intangible ones, come into play and characterise housing policy today; at least, this is the thesis we want to support in this article. The article's structure describes public housing in Rome and how the Corviale neighbourhood fits into this long, not always linear history. It then recounts the history of the Corviale neighbourhood, essentially its evolution over time and the processes of adaptation carried out by the residents, often as a rejection of the original project. It then describes the 'Quarto Piano' redevelopment project and its final conversion into flats, highlighting not only the structural aspects but also the social impact it has had. The chapter focuses precisely on these impacts by highlighting that today's housing issue represents a spectrum of situations far beyond the housing issue. From this point of view, the case study reproduced here presents the 'Laboratorio di Città Corviale' role in accompanying the redevelopment project. The laboratory, promoted by the University of Roma Tre, Department of Architecture, and supported by the Lazio Region, has played a role that should be known because of the results it has achieved in monitoring the implementation of the measures. The issue of housing, considered in the light of this case study, where we can not only look at the tradition but also the innovation of housing policy, and described in the final chapter, highlights some lines of action that can have a general value that goes far beyond the case reproduced here.

2. The public housing in Rome

From 1969 to 1989, the largest public housing project was carried out in Rome: 4456 ha of surface and about 700,000 rooms in 70 neighbourhoods; at the end of the program, in 1989, just over 500,000 had been built in 114 neighbourhoods [1]. Despite the shrinkage suffered, it remains the most extensive public housing project ever undertaken in Italy, twice as much as in Milan. The Plan for Economic and Social Housing in Rome (Peep) was prepared in 1964 and completed the first implementation phase in 1985 when the second Plan for Economic and Social Housing was approved, completing the implementation phase in 2006. The public sector intervention in social housing came in the wake of protests by the major industrial unions in 1969, which waged a critical struggle for the right to housing and against high rents. The general strike of 14 November 1969 remains one of Italy's most critical moments of social conflict in the entire post-war period. In this period of social struggle and conflict, marked by clashes in the streets with wounded and dead—a 'carabiniere' was killed during the general strike in Milan—negotiations began between the central government and the trade unions, leading two years later, in August, to Law 865/1971, the first (and still the only) organic law on housing in Italy. The law set in motion some legislative measures that it had already been prepared in the years before but had not yet been implemented due to lack of funds. Law 167 of 1962, for instance, was reissued and allowed municipalities to acquire land to construct neighbourhoods for economic and social housing [2]. The twenty-year golden age of public housing in Italy (1969–1989) significantly impacted the post-war design of the city of Rome. Much social housing and public facilities (schools, parks and gardens, car parks and churches, gymnasiums and health centres) were built in the 114 neighbourhoods. About 30% of all public facilities in the city of Rome are in these districts, which are home to only 12% of the capital's population. On average, there are about 43 square metres of public facilities per inhabitant, with a legal limit of 18 square metres. The neighbourhoods designed in the early 1970s were still committed to the models of the modern movement. They were designed as satellite towns to accommodate housing, services and public areas. They are housing estates ranging from a few hundred (500) to cities of 35,000. Initially, all areas were identified as urbanised ex-Novo, with most interventions taking place in the eastern part of Rome. The construction of the new neighbourhoods led to the urbanisation of large areas of land acquired through expropriation from public property and subsequently converted for housing, services and infrastructure. In the second quarter, on the other hand, more emphasis was placed on the restoration and reuse of the existing heritage. The new buildings were constructed in areas adjacent to buildings already built but lacking public services and facilities. These are Rome's self-built villages in response to the lack of public housing after the intense urbanisation that gripped Rome after the war. They are entire neighbourhoods for about 900 thousand inhabitants without basic infrastructure (sewage, water supply and public facilities). The villages were built outside the provisions of the master plan and are therefore considered illegal.

3. Corviale, a machine for living and adapting over time

The Corviale district is one of the most famous and well-known symbols in the production of public cities in Rome and beyond. It is part of the first Peep and is designed for about 6800 inhabitants, with public facilities of 50 square metres per inhabitant, far above the legal minimum of 18 square metres. Of the 60 hectares of land, 36% is used for services and only 7% for buildings. This is the most striking feature of the Corviale, a single urban building almost a kilometre long and nine stores high, with a street of shops and public services, professional studios and community spaces running through its interior. It acts as a barrier at the edge of the built-up city and faces west, towards the sea and the Roman landscape, the Valle dei Casali. This creates a scenario in which nature, agricultural and urban activities merge, forming an entirely modern landscape with the Roman countryside. A neighbourhood characterised by an extreme duality between density and rarefied has come to be seen as a radical architectural exercise (**Figure 1**). The kilometre-long residential building is counterpointed by a smaller building arranged at a 45° angle to the main building and crossed by an internal street that serves the shops and ends in the facilities connecting the different buildings and the different functions, the school, the market, the multipurpose areas and the sports facilities. Looking at the concept of a residential machine, it is undoubtedly the Roman quarter that comes closer to this theory than the others. However, the solution chosen by the planners, led by Mario Fiorentino, aims to refer to the characteristics of the site and reinforce the landscape dimension of the urban. The definition of the boundary, a double boundary, that of the building, closed in on itself, and that of the city, the last bulwark built to the west in front of the landscape, form the most important architectural feature of the neighbourhood (**Figure 2**). The dimensions of the building remain one of the most important urban signs of the city of Rome. The dimensions of the building are even more pronounced as they cause the features of the landscape. Here, the echo of a duality between the building and the emptiness of the landscape reinforces the dimensions of the one urban system—the building. Given the vastness of the emptiness of the landscape and the horizon, the house-city aqueduct stands out as a symbol of redemption for the working class who are the least of these—a popular house with a unique aesthetic.

Throughout its life, the building has undergone a series of signs and adjustments that represent a continuous deviation from the planners' intentions; in some cases, the residents reject the sophisticated architectural solutions. The management of a building of this complexity was not in the hands of the regional authority that manages public housing in Rome and Lazio (Ater), and the housing machine immediately yielded to the adjustments and changes that created space between the rooms abandoned to neglect, especially the non-residential rooms used temporarily for residential

Figure 1.
Corviale. Credit: Julian Schubert.

Figure 2.
Corviale. Credit: Laboratorio di Città Corviale.

purposes. The failure to open public facilities was the leading cause of the conversion of the fourth floor into temporary accommodation. This phenomenon can also be observed in other public housing neighbourhoods in Rome, for example, in Lurentino

38, where the 'bridge' buildings that were supposed to provide the neighbourhood with shops and offices were converted into makeshift shelters and then permanently occupied by families seeking housing. In Tor Sapienza, another public housing neighbourhood in eastern Rome, the 'service spine' at the centre of the residential courtyard has suffered the same fate: first decay and then 'abusive' conversion to residential use, and there, as in Corviale, a church is also one of the illegal uses. The particular architectural structure of Corviale, a single building almost a kilometre long, is cited as the main reason for the dysfunctionality of the neighbourhood and the degradation processes that have affected it. However, a comparison with the other neighbourhoods shows that the reasons also include the inadequacy of the public housing manager, the Ater (territorial housing agency). Ater had shown sensitivity and willingness to innovate in the planning and construction of neighbourhoods in the late 1970s, and then showed the inadequacies in the management of the housing, which were even more severe in the management of the non-residential spaces, in the phase of taking over the completed neighbourhoods and their management. These spaces, which are often an essential part of the spaces and buildings in the neighbourhoods, remained unused for many years immediately after construction. The newly completed buildings were looted; in some cases, doors, windows and bathroom facilities were removed. Looting that in many cases encouraged the process of squatting by individuals or groups of citizens who converted the spaces intended for non-residential purposes into makeshift shelters. In Corviale, the fourth floor of the building, the so-called 'piano libero', which was intended to accommodate all urban services, from shops to professional studios to community spaces and condominiums, suffered this fate before falling into disrepair and being abandoned and then occupied by those demanding housing and, consequently, adaptation to the housing. The shelters, built by the families themselves, re-used the spaces that were intended for non-residential purposes (**Figure 3**). The shelters, which became the destination of various housing routes, were connected by illegal routes. In other words, the houses did not follow the usual institutional procedure of public allocation. It was an actual illegal occupation. There was space, albeit not for residential purposes, and that was enough for individual residents and families to convert it for residential purposes to meet the need for social housing. The first inhabitants of the temporary shelters were

Figure 3.
'Quarto Piano'. Credit: Laboratorio di Città Corviale.

mainly Italian families, part of the migration that brought many people to the capital from the poorer areas of the south and the north of the country. In recent years, since the late nineties, the same makeshift shelters have been 'sold' to immigrant families from abroad, from the countries of Eastern Europe and North Africa.

The fourth floor, originally intended for services, was converted into living quarters, and thus resembled the other floors. The urban street inside the building thus lost all meaning as a place to meet and congregate. Over time, a process of adaptation has developed that is the result of the initiative of the individual residents and highlights their creativity in transforming the building.

The social project of coexistence, carefully thought out by the planners and taking shape in some spatial decisions, has thus been profoundly affected by a triple action—first abandonment, then appropriation and finally renewal.

3.1 'The important thing is that you can see the sunset from my house'[1]

The regeneration programme currently being implemented refers to the measures foreseen in the 'Contratto di Quartiere II' [3], which in the first phase foresees the demolition of about 130 self-built flats to be illegally converted into 'Quarto Piano' premises intended initially for services and shops [4]. In the second phase, the programme foresees the construction of 103 new social housing units for people entitled to social housing. The implementation of the measures envisaged in the regeneration project will be accompanied by the activities of the 'Laboratorio Città di Corviale', which was established in 2018 following an agreement between the Lazio Region's Directorate for Social Inclusion and the Faculty of Architecture of the University of Rome Three. The agreement implemented a city laboratories research project, activated with funding from the department, which envisaged bringing the university into the public residential areas of the city of Rome. The summer of 2018 saw the opening of the space, Ater has made available to the university for the laboratory's activities (**Figure 4**). The space is in the area of craft and market activities. Four years have passed since the first opening, and every Tuesday and Thursday, as well as many other days of the week, two researchers pull up the shutter of this space with the sign: 'Laboratorio di Città Corviale' (**Figure 5**). The activities of the

Figure 4.
The headquarters of 'Laboratorio di Città Corviale'. Credit: Laboratorio di Città Corviale.

[1] Cfr. www.laboratoriocorviale.it; anche Francesco Erbani, *Dove ricomincia la città*, Manni, 2021, Lecce pp.80-101.

Figure 5.
Sofia and Sara, the researchers. Credit: Laboratorio di Città Corviale.

laboratory concern social support actions related to the urban regeneration project of the neighbourhood promoted by the Lazio Region and Ater and the urban renewal action of the 'piano libero', the so-called 'Quarto Piano'. This is the project 'Il Chilometro Verde' [5], which consists of social housing construction where services were offered instead.

The designer Mario Fiorentino intended the 'Quarto Piano' for services, collective spaces, professional studios and urbanity. Those who lived on the upper or lower floors (the fourth floor also coincides with the change of building type from terraced house to the bannister's house) had to find themselves in a building in the city. The history of Corviale's 'Quarto piano' was different. Immediately after its construction and more and more over time, this space was adapted to the needs of a diverse population with different housing conditions, sometimes even with distant origins. These additional living spaces became inhabited and somehow became houses. The census had counted 135 households, but the regeneration project came up with 103 instead, a number lower than that of settled households but higher than those eligible for social housing. Families who had applied to the municipality of Rome's 'Bando Speciale' and met the requirements for social housing allocation are eligible for a new flat (73 applications of which 47 were accepted and 26 were not). The social support activity implemented by the 'Laboratorio di Città Corviale' aims to promote the 'Quarto Piano' regeneration project. This activity foresees the involvement of the current occupants of the illegal houses in the rehabilitation process. There are three moments in which the project will be carried out: the temporary departure of the occupants, for the time necessary to renovate the building, and then, only for those who have the requisites, the return to the new home. For those who do not have the right to return, the accommodation outside the neighbourhood remains and the temporary accommodation in one available Ater flats.

4. The 'Laboratorio di Città Corviale'

The 'Laboratorio di Città Corviale' qualified as an indispensable agency for implementing urban renewal. About half of the planned houses were built, and 70 families moved. The main steps were three: a thorough and direct knowledge of the inhabitants of the houses, with the identification of community leaders; people-to-people dialogue to explain the project, the timetable and the modalities of moving to other homes to allow demolition and reconstruction; and finally, the extension of intervention actions that affect the life of the neighbourhood and involve other protagonists who go beyond the issue of the house and imagine a future space that creates a heritage and roots. Work began in January 2019, and the site has continued in construction phases, according to the sequence described above. Occupants are being relocated, and spaces are being cleared for the planned demolition of the project's precarious artefacts and shelters (**Figures 6–8**). Ten phases are planned for the full implementation of the project, three of which have already been completed (**Figures 9** and **10**). Fifty-five families were affected, of which twenty-six had the right to be resettled as they were included in the ranking list. Twenty-five families who were not included in the ranking list have no right to return to the renovated shelters and remain in the emergency shelters provided by the Ater. In addition, four families were expelled because they did not meet the requirements for social housing. The lab's activities are diverse and cover a wide range, not limited to assisting the families who must be removed to carry out the work. There are coordination activities for the move, contacts with the new residents and support for tenants in their relationship with Ater technicians and officials, and also, meet with the affected families to get bureaucratic support for the move.

Figure 6.
The regeneration, work in progress. Credit: Laboratorio di Città Corviale.

Figure 7.
The regeneration, work in progress. Credit: Laboratorio di Città Corviale.

Figure 8.
Plan with illegal housing, before demolition. Credit: Ater.

Assist the families in finding the best possible placement for their children in the schools, and inform them about the phase of reintegration into the new flat, including the determination of the rent they will have to pay from the moment the new flat is handed over to them (**Figures 11–13**). In the meantime, in collaboration with local associations, the laboratory won a three-year tender for the 'Estate Romana'. It organises cultural initiatives in the neighbourhood and involves a community of artists

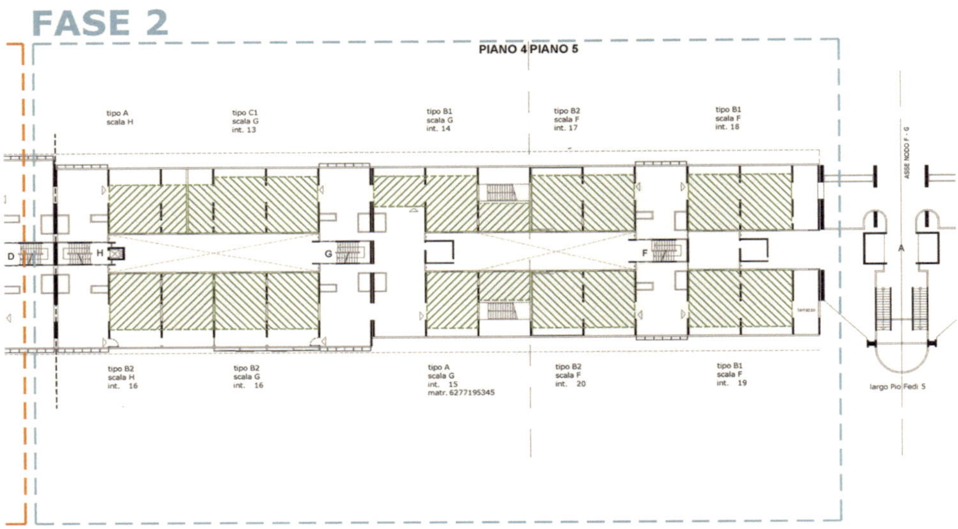

Figure 9.
'Quarto Piano', new flats, Phase 2. Credit: Ater.

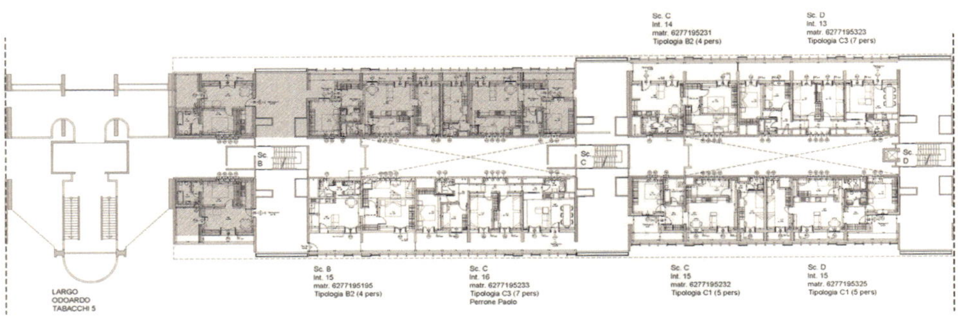

Figure 10.
'Quarto Piano', new flats, Phase 3. Credit: Ater.

Figure 11.
The new flat. Credit: Laboratorio di Città Corviale.

Figure 12.
The new flat. Credit: Laboratorio di Città Corviale.

Figure 13.
The new flat. Credit: Laboratorio di Città Corviale.

renewing the art and craft spaces they inhabit through the project 'Piazza delle Arti e dell'Artigianato' (**Figures 14–17**). A process called 'heritage making' has taken place, which is even more evident in the case of the 'Progetto delle Memorie', another laboratory activity. The 'Progetto delle Memorie' is a project that gives substance to the inhabitants' memories. It is the photographic and architectural inventory of the self-built houses on the 'Quarto Piano' before demolition (**Figures 18–22**). Photographs and house plans of the houses before demolition were exhibited in the spaces vacated by some of the houses, transforming the living hall of the first plot into a space for common use.

The 'Progetto delle Memorie' has given the inhabitants and all visitors a new and non-trivial view of a process of depositing housing that, even if it is outside the rules, cannot be considered marginal. The 'Quarto Piano' and the photos convey a sense of life that makes sense and dispels many of the prejudices often attributed to life in social housing.

An awareness that was then extended to the other inhabitants of the building by involving them in the 'Archivio Corviale' project, which collects photos of Corviale's first 40 years of life. The laboratory is a destination of many students from Italian and

Figure 14.
Social animation, Laboratorio di Città Corviale. Credit: Laboratorio di Città Corviale.

Figure 15.
Social animation, Laboratorio di Città Corviale. Credit: Laboratorio di Città Corviale.

non-Italian universities who stop for a few days to study what is happening in Corviale. It is happening that an idea of urban regeneration is taking shape where the centrality of the body located in space is being redefined, even renegotiated. It has initiated a particular intertwining in which the building project, the architectural one, the housing relocation, the rights and requirements of social housing, together with the right to education and health have given shape to an expression of living in the making that can only be carried out within the dimension of everyday life, of the gesture of care and taking charge. We have called this activity social accompaniment, where the emphasis on the word accompaniment indicates taking care of care as a posture in transforming the whole city and not just the physical space. An inhabitant pronounced the title of the paragraph during a dialogue with the researchers, who explained the transfer path, the demolition work and then the new house assigned to her. An expression denotes the assumption of a dimension of living that aspires to psycho-physical well-being that goes far beyond the mere satisfaction of an essential

DOI: http://dx.doi.org/10.5772/intechopen.107870

Figure 16.
Social animation, Laboratorio di Città Corviale. Credit: Laboratorio di Città Corviale.

Figure 17.
'Progetto, Piazzetta delle arti e mestieri'. Credit: Laboratorio di Città Corviale.

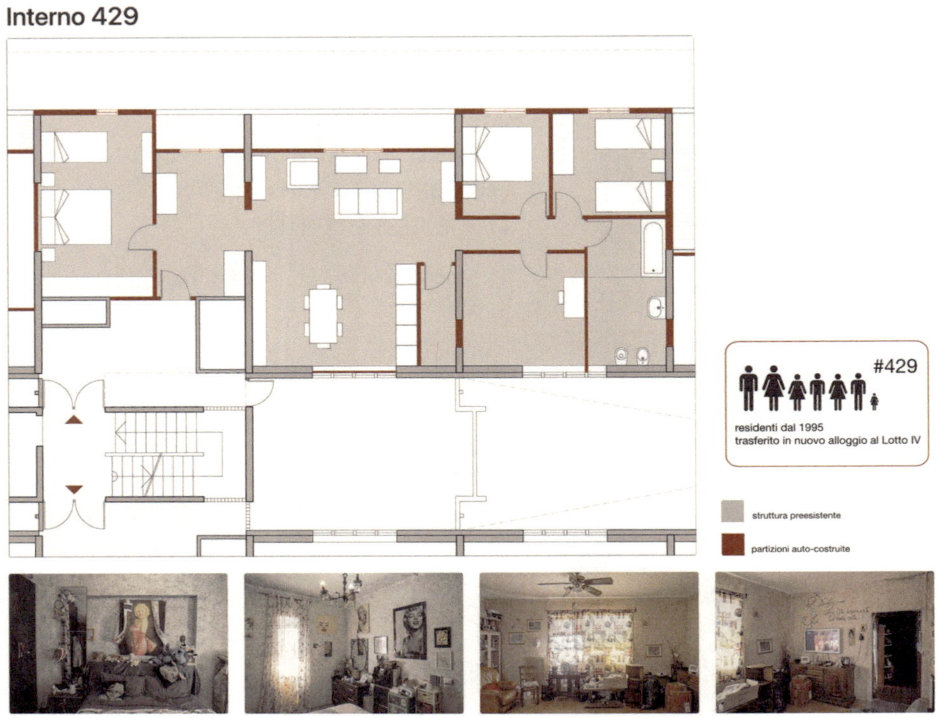

Figure 18.
'Progetto delle memorie', plan and pictures of the interior of the flat (Int.429) before the regeneration. Credit: Laboratorio di Città Corviale.

Figure 19.
'Progetto delle memorie', pictures of the interior of the flat (Int.306, 432, 434, and 438) before the regeneration. Credit: Laboratorio di Città Corviale.

need, having a house and a roof over one's head. Regenerating meant going beyond the basic needs, and this was possible thanks to the constant presence of the third party, the university, which has (re)built a bond of trust between institutions and

Figure 20.
'Progetto delle memorie', pictures of the interior of the flat (Int.306, 432, 434, and 438) before the regeneration. Credit: Laboratorio di Città Corviale.

Figure 21.
'Progetto delle memorie', pictures of the interior of the flat (Int.306, 432, 434, and 438) before the regeneration. Credit: Laboratorio di Città Corviale.

inhabitants. Interruptions of the construction site, bankruptcy of the construction company, and inaccuracies in the works presented as moments in which the construction of a common future were renegotiated. Things can also go wrong in urban transformation, but the lesson of the Corviale city laboratory is that regenerating can coincide with a change, even a radical one, a physical change but, above all, a change in the way of living in the city. The role that the Laboratory has built up over time inside the neighbourhood defines a series of activities. Social accompaniment in the relocation of the inhabitants, the social concierge as a mediator between the inhabitants and the Ater, cultural promotion with the 'il Progetto delle Memorie' and the transformation of the 'Piazzetta delle Arti e dell'artigianato' and, again, the project

Figure 22.
'Progetto delle memorie', pictures of the interior of the flat (Int.306, 432, 434, and 438) before the regeneration. Credit: Laboratorio di Città Corviale.

'Corviale si-cura' activated within the framework projects for the community provided by the ministry of Labour for citizens' income earners (**Figure 23**). The laboratory plays the role of weaving a plot around housing, working with inhabitants to take on the challenge of living together in a housing machine that has not only its rigidity but also shows an exciting ability to adapt to the deviations that life imposes on the designers' predictions. It is perhaps one of the most exciting aspects of how the neighbourhood changes, a co-evolution between the rigidity of the architectural system and the daily actions of the inhabitants who change its contents and realise the most appropriate ways of living for their needs.

5. Conclusion: it's not just a house

In Western society, the difficulty of accessing housing is one of the reasons why social inequalities among the urban population are growing; this also happens because public intervention in support of housing policies has gradually been reduced to the point of cancelling itself out [6]. Each European nation has prepared interventions

Figure 23.
Social project, Corviale si-cura. Credit: Laboratorio di Città Corviale.

that are partly different but are united everywhere by choice to sell public residential assets and by growing use of the market and intermediate forms, such as social housing or relatively affordable houses to cope with the demand from the middle classes. Direct public policies that provide for the construction of housing for the less well-off social classes are lacking or at least marginal. In Italy, a further complication has been added; following the process of federal devolution, housing policies have been entrusted to the regions, which in Italy are twenty. Each region has produced at different times and with very different references, choices and in some cases, very different organic housing policies, and it is not easy to return a unitary picture of public action in Italy. Overall, the latest data on the construction of public residential housing in Italy stops in the order of a few thousand; the 1993 data, at the end of the twenty-year golden cycle of public construction in Italy, were about 40 thousand new accommodations per year. In this context, the Corviale regeneration intervention with the replacement of self-built housing by the 'squatters' represents a unique case that indicates some lines of action for possible interventions.

Meanwhile, the regeneration processes may involve the transformation of environments already built through the change of intended use from services or offices or more to public residences to be rented to the neediest families. Even if it is about houses, what comes into play is much more than just a house. It is a readjustment process involving urban and social impacts, even when it has a predominantly building character. The mechanisms underlying the regeneration processes necessarily involve the inhabitants and institutions in a dialogue, which, as demonstrated by the activity of the 'Laboratorio di Città Corviale', constitutes the prerequisite for the regeneration process to achieve its aims and be successful.

A second lesson given by Corviale is that we can go back to designing in public neighbourhoods, a legacy of the modern movement, and find there, in those neighbourhoods, new possibilities for densification or rather intensification, in the use of the already built space that is often abandoned and underused. It is a space of action for the urban project connected to the housing demand that is also particularly important for the design results that may involve redesigning some of the principles with which those neighbourhoods were conceived and built. In some cases, fifty years after their construction, it is now possible to imagine a rehabilitation intervention that represents a new urban stratification that considers the adaptation processes that the inhabitants have brought into these architectures over time, in some cases, real machines for living. A third lesson concerns the forms of living, end of a housing model, which was concerned only with responding to the basic needs of a shelter house, has long since been sanctioned. Today, even in public housing, it is essential to focus not only on the house but on the model of living, knowing how to identify the neighbourhood even before the space of the house. In common space, necessary for the sociability of being together is presented as the place for constructing the public dimension alongside that of the private, single and individual dimensions. This alternation of rhythms and times configured in diversified spaces and housing solutions is now unavoidable in every housing project.

Moreover, finally, there is one last lesson: the processes of regeneration are accompanied by an activity of roots, which we can also call identity, which passes through the production of heritage. It is an activity that involves different aspects. Important among them are those that also involve spaces, the construction of collective memory, and the layering of signs due to cultural events intended to leave a mark on the neighbourhood's lifestyle. In the 'Laboratorio di Città Corviale' activity, participation in tenders for the Roman summer or the promotion of social and cultural integration activities has been a decisive factor in the success of the laboratory's activity and in achieving the objectives pursued by urban regeneration interventions. Ultimately, it is an agency activity that starting from the inadequacies of public institutions, especially regarding housing in sensitive neighbourhoods, carries out an important activity to promote the possibilities of the neighbourhood's physical, economic and social transformation, a role that still focuses on dialogue with the inhabitants and their direct participation in mediation. The protagonist of the inhabitants is not a convenience of the institutions, the tenders or the projects financed within the framework of specific projects represent a constant presence in the neighbourhood of this accompanying activity. 'Il Laboratorio di Città Corviale', set up by Roma Tre University, represents a case study to understand the new housing demand and the new lines of action to create public housing and accompany the physical transformation of neighbourhoods.

In conclusion, we can say that the four-year experience of regenerating Corviale, which is still ongoing, can be an important case study in the field of housing policy and public policy to observe to guide the future of housing policy. Particularly in terms of interventions in the already existing, in the already built city, where the goal is undoubtedly to make the right to a house accessible, but where the integrative dimension of urban policy also comes into play, it is about giving a house, but let us talk about much more than the individual house, it is the right to the city.

Acknowledgements

Thanks to the Lazio Region, Social Inclusion Directorate in the person of Dr. Antonio Mazzarotto and his collaborators Tonino Sammarone, Luca Piras and Rosario Mancuso. Thanks to Dr. Sofia Sebastianelli, who accompanied the research project of the 'Laboratori di Città' from the beginning, and to the whole territorial team of the 'Laboratorio di Città Corviale', Sara Braschi, Maria Rocco, Sara Le Xuan for the commitment they put into this rich and exciting experience from many points of view. Thanks to colleague Francesco Careri.

Conflict of interest

The authors declare no conflict of interest.

Author details

Giovanni Caudo†
Architecture Department, University of Roma Tre, Rome, Italy

*Address all correspondence to: giovanni.caudo@uniroma3.it

† The "Laboratorio di Città Corviale" has been promoted by the Department of Architecture of Roma Tre University and supported by the Department of Social Policy of the Lazio Region. The scientific coordination of the laboratory is carried out by Prof. Giovanni Caudo, urban architect, and Prof. Francesco Careri, architect. The territorial team is composed of researchers: Sofia Sebastianelli, Sara Braschi, Maria Rocco, Sara Le Xuan. The laboratory uses external collaborations and has started a collaboration with Avanzi spa.

References

[1] Giovanni Caudo, Mauro Baioni, Nicola Vazzoler. Il suolo conteso: intenzioni e sopravvenienze nel secondo Peep di Roma. Territorio. 2019;**90**:37-45

[2] Paola Di Biagi, La città pubblica. Edilizia sociale e riqualificazione urbana a Torino, Allemandi, Torino. 2008

[3] Sofia Sebastianelli, Sara Braschi, Il Pubblico Re-Esiste in periferia. L'esperienza del Laboratorio di Città Corviale, in Anna Delega, Elisabetta Ginelli (a cura di), Storie di quartieri pubblici: Progetti e sperimentazioni per valorizzare l'abitare, Mimemis edizioni. 2022:278-285

[4] Francesco Bargiggia e Massimo Bricocoli, Politiche per la casa e strumenti di riqualificazione urbana. I contratti di quartiere in Lombardia, Edicom Edizioni, Milano, 2005

[5] Mura G. A che punto è il Kilometro Verde di Corviale a Roma? Parola a Guendalina Salimei. In: Artribune. 2021. Available from: https://www.artribune.com/progettazione/architettura/2021/01/corviale-architetti-guendalina-salimei/

[6] Gallent N. Whose Housing Crisis. London: Policy Press; 2019

Chapter 7

Perspective Chapter: Reimaging Affordable Housing through Adaptive Reuse of Built Heritage

Sasha Tsenkova

Abstract

This chapter focuses on adaptive reuse of heritage for affordable housing in Canadian cities. The issue is critical in the context of efforts to create socially inclusive and affordable cities through integrated urban planning, heritage conservation and housing policies. The research has three main components. First, it provides a framework for future urban regeneration emphasising the environmental, economic and social aspects of sustainability. Second, it reviews the synergies between adaptive reuse and affordable housing provision and provides a compelling rationale for their integration. Finally, it outlines three main approaches to adaptive reuse—typological, technical and strategic—arguing for implementation through 'policy-planning-partnership' nexus. Using illustrations from successful affordable housing projects through adaptive reuse, the research demonstrates the importance of urban regeneration where strategic investment in diverse, socially cohesive affordable housing sustains the vibrancy and vitality of inner-city neighbourhoods.

Keywords: adaptive reuse, heritage conservation, sustainable urban regeneration, affordable housing, Canadian cities

1. Introduction

Cities are recognised as divers of the urban economy and centres of innovation, but experience a persistent shortage of affordable housing and growing social inequalities that affect the health and well-being of urban communities [1]. In the context of fiscal austerity and global inflation, the growing shortage of affordable housing in Canadian cities is affected by changing politics, housing market inefficiencies and concentration of urban poverty. Regardless of important historic and economic differences across Canadian cities, the recent pandemic exacerbated existing social inequalities and social exclusion [2] and demonstrated the importance of affordable housing. People need a place to call home more than ever, a place that provides shelter from economic and social stress, a place to live, work, educate the children, care for family members and maintain public safety through social distancing. Governments during recent lockdowns provided emergency shelters for the homeless, extended mortgage payments, introduced rent deferral and other emergency measures to temporarily shelter people from immense housing difficulties and protect the public [3]. The unprecedented challenges to public health in cities have demonstrated the need to consider affordable housing as a critical part of social infrastructure that requires sustained long-term investment and support to establish a resilient ecosystem. Social equity has emerged as a key urban policy, including the need for strategic transformation of our built environments using principles of social, environmental and economic sustainability [4].

Within this context, the research addresses a vital area for urban planning that can contribute to more inclusive and equitable cities. It identifies a solution to the affordable housing crisis through adaptive reuse of heritage buildings in inner-city communities. The focus is on the experience of Canadian cities, home to over 80% of the people in Canada, where the growing shortage of affordable housing has prompted urgent action by all three levels of government. Recognising the environmental, economic and social synergies of adaptive reuse of heritage for affordable housing, the research outlines a framework for sustainable neighbourhood regeneration. Using insights from different projects, it offers a blueprint for diverse implementation at different scales—from project-based intervention to strategic neighbourhood regeneration through integrated programs and partnerships. The research methodology includes literature review of studies on affordable housing and adaptive reuse to develop a conceptual framework for sustainable heritage conservation strategies. This is complemented with a nested case study method, combining analysis of projects in Canadian cities to illustrate patterns of diversity. Given the importance of sustainability to adapt built heritage for affordable housing, the methodology also includes content analysis of planning and policy documents pertaining to the research focus, key informant interviews and visits to project sites.

The research approach draws on housing and heritage planning studies. These two streams provide opportunity to connect diverse policy perspectives to planning and urban design aspects critical for the advancement of social sustainability in the city [5, 6]. The research views built heritage as a spatial arena of adjustment through adaptive reuse where sustainability planning and design generate positive outcomes for people and historic places. Insights from successful projects highlight possible synergies and partnerships to address both the lack of affordable housing and the loss of historic and cultural heritage in Canadian cities. Heritage conservation and housing share a strong synergistic tie that underscores the importance of urban social sustainability [7, 8].

Adaptive reuse is a process that converts heritage buildings to new use, maximising the economic and social benefits of heritage structures while restoring their value to a community [9]. It provides a physical link to the past social, economic and cultural development of a place, retaining a sense of the previous form while providing opportunities for the future [10]. Through adaptive reuse of heritage planners have the opportunity to address displacement in low-income inner-city communities, contribute to larger community well-being and sense of place. The strategy proposed in this research capitalises on the untapped potential of adaptive reuse of historic buildings for non-market housing with a social purpose. These tangible assets of our cultural heritage have become physically or functionally obsolete due to rapid societal changes in technology, standards and local economies. Most obsolete historic buildings are either demolished, resulting in a loss of over 20% of Canadian heritage, or converted into museums, luxurious apartments and entertainment complexes as heritage are often perceived as a commodity [6, 11]. It is essential to reimagine these places in a creative way to increase the small share of non-market, socially owned housing, which is less than 6% in Canada. Due to the front-end loaded nature of housing costs, the process is dependent on a consistent alignment of a range of financial and regulatory instruments, such as cost-sharing government programs that subsidise the capital-intensive conversion as well as a variety of planning incentives (e.g. inclusionary zoning, reduced land costs and taxes, alternative standards) to incentivise development and heritage conservation.

2. Reimagining affordable housing: sustainability of adaptive reuse

Concepts of sustainability have gained significant ground in city building, emphasising a more holistic approach to urban regeneration practice. Achieving sustainability in existing neighbourhoods is a long-term, complex process of conflict resolution among environmental, economic and social sustainability, implemented through the lens of planning and housing policy integration.

2.1 Environmental sustainability

As Canadian cities face the increasing global impacts of climate change, ongoing environmental degradation and higher energy costs, planners and policymakers have placed significant emphasis on addressing environmental sustainability in the built environment. The physical form and location of built heritage represent a key opportunity for cities to make progress towards environmental targets by focusing on reuse, recycling and redevelopment of existing housing. This supports circular economy strategies that minimise environmental impact by extending the use of materials and reducing the consumption and waste of materials and energy. Retrofitting and adaptively reusing existing heritage buildings for housing purposes represents a significant opportunity to promote more sustainable uses of renewable and non-renewable resources. Circular economy strategies call for a cultural shift within the construction and development industries, to view buildings as reusable resources as opposed to consumable products to achieve desired sustainability outcomes [12, 13].

Notwithstanding the progress made on the technical side to increase the sustainability and energy efficiency of existing housing through energy efficiency retrofits, tensions often arise between building preservation and conforming to current regulatory requirements. While many current building and energy codes emphasise more environmentally sustainable building practices, retroactively updating older buildings to these standards can act as a barrier to building reuse and threatens the financial viability of a project. In light of these tensions, and additional factors around density, location and building condition, studies have found that incentives and flexibility in planning and housing policy administration are needed to ensure retrofits and adaptive reuse projects in existing heritage buildings are successful and desirable [13, 14].

2.2 Economic sustainability

It comes as no surprise that in urban regeneration projects, economic sustainability and viability are critical, influencing social and environmental sustainability performance, in addition to the decision to demolish or reuse. Costs of retrofits, property taxes, financing and rate of return on investments may result in rehabilitation costs being higher than new construction. While these factors can act as economic barriers, they also pose an opportunity for regulatory bodies to incentivise and facilitate the economic sustainability of housing regeneration and retrofits. Currently, government incentives, tax credits and housing policies often act independently as opposed to synergistic tools to support adaptive reuse and broader sustainability goals [14, 15].

A major barrier to urban regeneration and adaptive reuse is the short-term, capital-intensive investment required to achieve long-term sustainability outcomes. Collaborative efforts are required to realise these complex, multi-stakeholder projects, and these collaborations are reliant on sustained financial support to be successful [1, 16]. The availability of public funding and low-cost finance to the private industry not only helps reduce the initial financial barrier to adaptive reuse but also helps mitigate the prevalent risk of unexpected challenges and costs. Another important benefit from reusing and retrofitting individual housing is the contribution to economic, social and environmental value in the surrounding neighbourhood. Looking beyond the market value of a building is essential to maximise neighbourhood scale effects, considering externalities and spillover effects in the surrounding urban environment. In addition to job creation, improvements in housing quality and neighbourhood services, urban regeneration of heritage through its adaptive reuse is a means to achieve smart growth principles, encouraging compact development, use of existing infrastructure and higher densities [1, 4]. Developers and owners must balance the economic benefit of developing a site to its highest development potential, with the high opportunity cost forgone for building retrofits and conservation. Studies call for a broader approach to analysing return on investment that includes social and environmental qualities to adequately measure the community-scale benefits of urban regeneration and adaptive reuse [13, 15].

Through the planning and design stages of urban regeneration projects, planners have the opportunity to make a significant difference in environmental and economic outcomes for development. As a strategic process, planning for adaptive reuse is an effective tool for intensification, redevelopment and provision of affordable housing in existing neighbourhoods. In comparison to the more rigid factors of capital investment and asset condition, which define the feasibility of retrofits and adaptive reuse at the project level, regulation and policy created through planning are relatively dynamic, providing a key opportunity to enable larger-scale transformation and synergy [10, 14].

While a lack of resources to incentivise and regulate private industry and individual homeowners in Canadian cities can be a barrier to achieving sustainability, planners and municipal authorities are starting to address these issues on a larger scale, focusing on neighbourhood effects and affordable housing. Land use planning plays a critical role in the delivery of place-based outcomes such as complete, mixed-use communities that go beyond single-use zones to offer a diverse range and mix of housing options, densities and tenure [2, 16]. Zoning is a powerful land use tool, directly impacting the form, use, scale, occupancy and other aspects of our cities, including the sustainability and viability of adaptive reuse projects [17].

2.3 Social sustainability

While significant attention in practice has focused on the environmental and economic benefits of urban regeneration through adaptive reuse, social sustainability has been undermined. Yet, social sustainability is critical as the process presents a valuable way to generate better places, boost economic development and preserve built heritage but also incentivises communities to embrace more sustainable lifestyles [10]. In an increasingly privatised, neoliberal city, challenges to integrating social sustainability principles in the planning and regeneration of built heritage in Canadian cities are significant. These places have diverse and historically defined urban forms as both people shape the places they work, live and play, and places in turn shape their inhabitants. This diversity requires a context-specific approach to urban regeneration, preservation and adaptive reuse so that it contributes to larger community well-being and sense of place. Despite many common features defining the urban form and physical characteristics of built heritage in Canadian cities that have a relatively short history, it is important to recognise the cultural and social diversity of the people living in these areas, their lifestyles and traditions that established the character, landscape and history of these urban areas. The 'one size fits all approach' is not feasible.

Quality affordable housing is a key component of social sustainability, in the form of mixed-income, and mixed tenure neighbourhoods [2, 18]. Older buildings provide a significant source of affordable housing. With a focus on social sustainability, urban regeneration projects in heritage districts and existing neighbourhoods can leverage the redevelopment processes as an opportunity to produce below-market-rate housing, creating positive impacts for many vulnerable people living in post-socialist cities. Furthermore, the degree of social sustainability of any retrofit and adaptive reuse project is highly dependent on accessibility to essential services, both at a neighbourhood and project scale. Adaptive reuse strategies that focus on the needs of underserved and vulnerable communities can support greater accessibility of the public realm, transit and social services such as education and health care, while helping counter the impacts of gentrification [14]. Preserving and enhancing the urban block structure and walkability in conjunction with active transportation systems contributes to the affordability and accessibility of housing that serves the needs of low-income families and seniors [15].

The role of participatory planning processes is a necessary component of achieving social sustainability. Within a holistic planning framework for urban regeneration through adaptive reuse, it is important to collaboratively engage with the community through transparent and collaborative processes to determine priorities in local needs, mobilise local support and leverage financial and in-kind contributions to various aspects of the implementation.

3. Heritage and affordable housing: synergy in adaptive reuse

The literature on heritage conservation, adaptive reuse, affordable housing, historic buildings conservation and revitalisation addresses the synergy between these sectors in three domains—environmental, economic and social.

3.1 Environmental synergy

The original construction of obsolete buildings, including disused or underused historic buildings, expended a large amount of embodied energy through material extraction, production and transportation. Reuse of such structures is also the reuse of embodied energy and hence, avoiding demolition waste and reducing the amount of energy consumption [10]. Adaptation of obsolete historic buildings is also an opportunity to incorporate energy-efficient features to bring old structures in line with current building legislation in terms of safety and sustainability. The environmental synergy between the two sectors not only conserves energy as the 'greenest building is the one already built'

The social housing is adjacent to the boardwalk in the Beaches, one of the most attractive historic neighbourhoods in downtown Toronto, and is itself a historic resource. The building contributes to the character of the place and provides an opportunity to integrate social housing tenants in a high-income area. During the reconstruction, only the original façade of the 3-story, the 27-unit property was preserved. In addition to the installation of an elevator and other accessibility features, the primary goal of the regeneration was to meet a 40% energy efficiency improvement. The total cost of the regeneration was $5,894,340, and it provided 27 apartments of mixed-income housing. Funding from energy efficiency retrofit and neighbourhood environmental programs covered project costs. Half of the original tenants were able to come back to Hubbard Boulevard and live in rent-geared-to-income units where their pay 25% of their income in rent [16].

Box 1.
42 Hubbard Boulevard, Toronto energy efficiency by design.

but also reduces the ongoing housing costs in the social housing sector through energy-efficient retrofits and improvements in the quality of technical systems and services [16]. Rehabilitation of historic buildings into affordable housing is an innovative way to recycle the expended material and energy and reduce housing costs for low-income households and social housing providers as the project featured in **Box 1** illustrates.

3.2 Economic synergy

Both affordable housing provision and historic buildings preservation are areas that require government intervention as the market on its own are unable to respond in an adequate way. Many governments provide funding or incentives to encourage affordable housing and historic building rehabilitation projects. Tax incentives, grants and ongoing rebates for retrofits have been successful at encouraging developers to pursue projects that convert obsolete historic buildings into affordable housing in the USA and many European countries [4, 5]. Such financial and fiscal support is essential in ensuring the feasibility of affordable housing projects, which usually do not have high return on investment, as well as incentives for rehabilitation projects that have high risk of cost overrun due to unforeseen technical complications [19, 20]. Public funding for such projects also leverages private investment and philanthropic donations. The nature of such projects is likely to attract financial contribution from non-profit organisations and affluent individuals who believe in both causes. This synergy allows for a more diverse funding envelope, where developers can maximise the potential of unique heritage assets, and infuse a mix of uses—retail, arts, culture—to complement housing and create attractive places. Local developers can obtain available funding for both sectors to increase the economic viability of these projects (see **Box 2**).

Formerly home to Imperial Oil's Saskatchewan headquarters and a national bank, the Derrick building was converted into Renaissance Retirement Residence, a 164-unit senior housing in downtown Regina. The new building has a variety of unit sizes and targets low- to moderate-income seniors. The adaptive reuse incorporated environmentally sustainable features such as geothermal heating and cooling, solar-heated domestic hot water, a waste heat recovery system and an energy efficient building envelope. Additional floors were added to provide more space for community amenities. The project was financed by all three levels of governments through a combination of financial and fiscal support as well as the private developer [21].

Box 2.
Renaissance retirement residence in Regina.

In Canada, the National Housing Co-investment fund pledged about $16 billion in funding and low-interest loans for the creation of 60,000 affordable housing units while historic building preservation projects could seek funding from cost-sharing program at the provincial level. Provinces provide funding to conservation projects for designated historic buildings [6]. Despite a more limited scope compared to other countries in Europe and the US, the existing funding in Canada offers a favourable environment for such synergistic projects.

3.3 Social synergy

Many obsolete historic buildings are found in or in close proximity to city centres [6]. In many Canadian cities, the majority of historic buildings are located in city centres: approximately 90% in Toronto, 82% in Calgary and 45% in Vancouver. Downtown revitalisation efforts have resulted in the concentration of jobs, retail and good access to public transit services. Similarly, public services that provide support to low-income and homeless people such as food banks, housing help centres and social services providers are heavily concentrated in downtowns of major Canadian cities.

Inner-city neighbourhoods have been plagued by non-descript high-rise towers and unwelcoming 'projects' that were a manifestation of the misguided urban renewal initiative [2]. The most successful neighbourhood revitalisation efforts were attributed to the retention and reinvestment in the historic fabric of the neighbourhood, retaining its legible urban blocks, walkability and gentle density. The recognition of the character and quality of historic buildings and their adaptive reuse further contributed to the legacy of the place and its uniqueness. The scale and unique architecture of historic buildings make them ideal for conversion to mixed-income housing with a low-income component [14, 18]. Leveraging this synergy provides an effective method to address one of the housing market inefficiencies—the inability to effectively allocate land to various actors based on needs rather than financial capacity. As such, converting historic buildings into affordable housing through adaptive reuse provides access for low-income groups to cultural heritage resources and high-quality environments (see **Box 3**).

Coxwell Stables in Toronto is an example of a small-scale adaptive reuse project. Originally built in 1919 for the horses that pulled the Toronto Public Works Department vehicles, the site was designated by the Toronto Historical Board in 1981, and was bought by the City of Toronto's non-profit housing corporation (Leslieville Historical Society, 2020). The adaptive reuse of Coxwell Stables provided 11 affordable housing units while retaining the historic character of the building. The cost per unit was less than $100,000 and the project received financial support from the provincial affordable housing programs and local government grants for historic places [22]. *Coxwell Stables Redevelopment*

Box 3.
Coxwell stables in Toronto.

Existing neighbourhoods with obsolete historic buildings often have lower property values. Adaptive reuse is contributing to the effective redevelopment of existing assets but can provide affordable housing opportunities to target groups that might not be able to find a suitable alternative in the mainstream housing market. Single-person households and people with special needs might benefit from the locational advantages of downtown services, transit and jobs (see **Box 4**). The rehabilitation of the underutilised buildings will help revitalise the immediate neighbourhood and

The Heritage block at 18 West Hastings Street in Vancouver is a six-storey brick Edwardian building built in 1909. Renovated by Reliance Properties and ITC Construction Group it provides 30 suites in Vancouver's Downtown Eastside with the concept of contemporary "micro-loft". It has the smallest self-contained rental apartments of approximately 270 square feet for single people. The adaptive reuse project deploys unique design strategies to retain character-defining elements and the integrity of the built structure, while providing flexible layouts and floor plans for the affordable housing units [23].

Box 4.
Micro lofts in downtown Vancouver.

generally receives less opposition from the local community as any use is better than abandonment. This is especially helpful for affordable housing projects, which usually receive high level of pushback from the local residents [24].

4. Adaptive reuse of heritage for affordable housing: planning-policy-partnership nexus

Understanding the multiple perspectives of adaptive reuse of heritage for affordable housing requires comprehensive planning and coordination of design intervention. It makes the case for an integration of different theoretical approaches to adaptive reuse and heritage conservation—typological, technical and strategic [25]. These dimensions are presented in **Figure 1**.

4.1 Approaches to adaptive reuse

In the Canadian context, cultural heritage formally becomes a historic place when an authority recognises its "heritage value and character-defining elements" ([26], p. viii). An understanding of values is essential for successful heritage conservation and forms the basis for adaptive reuse projects [6, 25]. The typological approach focuses on building use and function as a primary determinant for adaptive reuse, categorising the barriers and success factors towards reuse by each building typology's

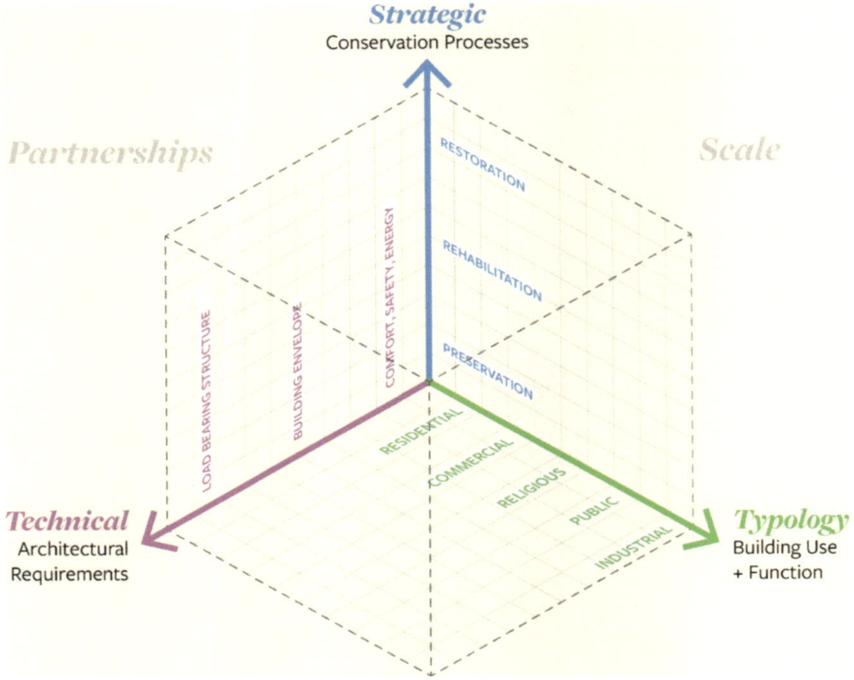

Figure 1.
Synergy of adaptive reuse approaches.

historical or contemporary use. High-level classification of building typologies includes industrial, religious, (semi-)public, residential, military and commercial uses [25, 27]. In practice, the typological approach to adaptive reuse is regulated at the municipal level through zoning and land use. These regulations often control other aspects of design in addition to use, including site coverage, floor area ratios, height limits, building envelopes, relationships to adjacent buildings, and parking requirements, all of which influence the feasibility and success of adaptive reuse projects ([6], p. 122). Ensuring flexibility in zoning and application of relevant building and fire codes with adaptive reuse projects can serve the nuance of typology-specific requirements and support adaptive reuse from historic functions to contemporary uses. In affordable housing projects, the building use is often mixed, combining a range of opportunities to meet contemporary needs beyond residential.

The technical approach focuses primarily on the technical aspects required to reuse a building, providing guidance and discussion on upgrades to the load-bearing structure, the building envelope, and the comfort, safety and energy efficiency of the adaptive reuse project [25]. Developed through a primarily architectural and engineering lens, the technical approach has manifested itself in the creation of guidebooks and technical expertise on the physical 'how' of adapting a building to allow new functions. In Canada, adaptive reuse is underpinned by the impacts of conservation efforts on a structure's 'character-defining elements'. While not a technical document, the *Standards & Guidelines* provide a philosophical approach and framework that reinforces technical decisions that impact built cultural heritage [26].

The strategic approach focuses on the analysis of the tangible processes and strategies required for the architectural conversion of heritage buildings. These strategies are physical interventions to convert the building to new uses, providing the guiding design concept. The original building is considered critical to the adaptive reuse design strategy [27, 28]. In the Canadian context, the overarching strategy for protecting historic places is conservation, defined as all actions or processes that are aimed at safeguarding the character-defining elements of a historic place to retain its heritage value and extend its physical life. The three strategies for heritage conservation are preservation, rehabilitation and restoration, or any combination of these actions or processes.

4.2 Partnerships and scale of adaptive reuse

The adaptive reuse of heritage for affordable housing due to its social complexity requires a creative approach, which aims to initiate a plan of action that creates synergies between the typological, technical and strategic approaches with the program requirements that provide the fiscal, financial and planning/regulatory support for social housing. Following an initial phase of investigating the existing building and understanding the project's design philosophy, the decision of whether to preserve, restore and rehabilitate is made [27, 28]. In addition to the synergies of the typological, technical and strategic approaches to heritage reuse, the scale of planning and development, and the formation of partnerships are key determinants to a holistic adaptive reuse approach in Canada (**Figure 1**).

With limited funding, resources and regulatory tools available at the federal level in Canada, "working in collaboration and in partnership is essential to ensure cultural resources at heritage places are safeguarded" ([28], p. 21). Partnerships are critical to incentivise private development, balancing the cultural and social sustainability goals with the profit-oriented motivations of the private sector. Partnerships that integrate the community are an important tool to bridge the gap between limited public financial resources, and the need to revitalise urban areas. These partnerships not only ensure successful project outcomes but create the foundations of knowledge networks to facilitate future small- and large-scale projects. Moving towards a system of 'less policy – more partnerships' can help alleviate risk and create opportunities where the private sector would otherwise choose more economical alternatives [1, 2].

The scale at which adaptive reuse projects occur is not limited to individual buildings but can vary greatly across scales from facades, to buildings, to groups of buildings and heritage districts. A multi-scalar approach to adaptive reuse considers the implications and challenges present at various levels, understands the inherent complexity of adaptive reuse projects and generates robust solutions to address this

> Once the workspace for 12,000 workers of the Canadian Pacific Railway, the 50-hectare site was abandoned in 1992, leaving thousands of local residents without jobs. The site's original redevelopment into a shopping mall met with opposition from the local community that demanded the preservation of the industrial heritage and the inclusion of social housing in the redevelopment project. The first phase saw Angus Shops transformed into a mixed-income housing complex of 2587 units with 40% social housing targeted at low to moderate-income seniors and families. Subsequent phases added retail and job opportunities as well as residential units with 20% social housing for local residents. Overall, the regeneration was successful at lowering the unemployment rate of the neighbourhood and infusing a large number of social housing—cooperative, affordable rental and subsidised ownership. However, the neighbourhood regeneration also brought in an influx of new condominiums and businesses that triggered gentrification [29].

Box 5.
Adaptive reuse of heritage for affordable housing, Angus shops redevelopment in Montreal.

complexity at the building and urban scales. At the same time, the move towards neighbourhood-based regeneration requires the scaling up of approaches usually implemented in individual buildings to generate sustainability outcomes for people and historic places as the example in Montreal indicates (see **Box 5**).

4.3 Planning-policy-partnership nexus

An important strategy for change that might be successful in the sustainable regeneration of inner-city neighbourhoods in Canadian cities builds on the planning-policy-partnership nexus [25, 26]. Nexus thinking transcends traditional policy and decision-making silos and develops approaches that build synergies across these sectors. Partnerships for affordable housing in cities and neighbourhood revitalisation are indeed very diverse multi-sectoral collaborations that leverage real estate market pressures to promote affordability goals and social mix. Cities often take the lead in managing the planning-design-policy nexus as neighbourhood rebuilding takes decades and shifting the responsibility to private developers might not work, particularly in the context of gentrification and displacement of lower-income residents. Partnerships need robust and sustained financial support, alignment of planning policies and institutional commitment to increase the supply of affordable rental housing. Such complexity by design makes statements on 'what works' and 'what does not' challenging and illustrates the interdependent nature of resilience at the nexus, raising the fundamental questions of how policy might enable systemic resilience [1]. Each city will need to develop its own successful model, based on the resilience of the planning-design-policy nexus for affordable housing to respond to growing affordability pressures while emphasising diversity and social mix [3, 28].

The experience of major Canadian cities in the context of urban regeneration illustrates opportunities for synergies of different policy frameworks guiding heritage conservation and provision of affordable housing. This research has demonstrated the potential of integrated approaches to adaptive reuse at the project/building scale as well as more strategic area-based action planning to generate a wider range of positive outcomes associated with such projects. Ultimately, the efficiency and effectiveness of heritage conservation through adaptive reuse and alignment of neighbourhood sustainability goals is enhanced as the whole is greater than the sum of its parts. Affordable housing partnership models in Canadian cities offer one possible solution to a growing affordability crisis adding adaptive reuse of heritage to their planning-policy-design strategy toolbox repertoire. Such synergy allows the scaling up of limited project/building-based experiences to a more strategic level, emphasising the importance of socially diverse communities with jobs, opportunities and services that have a unique historic identity and sense of place [30].

Acknowledgements

This chapter builds on research supported by the Alberta Heritage Resources Foundation in Canada. I am grateful to my research assistants—Jeremy Tran and Jay Heule for their commitment and rigorous collaboration in the review of the literature on adaptive reuse of heritage.

Author details

Sasha Tsenkova
School of Architecture, Planning and Landscape, University of Calgary, Calgary, Canada

*Address all correspondence to: tsenkova@ucalgary.ca

References

[1] Tsenkova S. Affordable housing and the future of cities. In: Tsenkova S, editor. Cities and Affordable Housing. Planning, Design and Policy Nexus. New York: Routledge; 2021. pp. 1-12

[2] Tsenkova S. Neighbourhood rebuilding and affordable housing in Canadian cities. Urban Research & Practice. 2022. pp. 773-778. DOI: 10.1080/17535069.2022.2082023

[3] Moraci F, Errigo MF, Fazia C, Campisi T, Castelli F. Cities under pressure: Strategies and tools to face climate change and pandemic. Sustainability. 2020;**12**:7743. DOI: 10.3390/su12187743

[4] Krapp M, Vaché M, Egner B, Schulze K, Thomas S. Housing Policies in the European Union. Darmstadt: Institut Wohnen Und Umwelt GmbH, IWU & Political Science of Technical University Darmstadt; 2020

[5] Scanlon K, Whitehead C, Arrigoitia M, editors. Social Housing in Europe. Oxford: Wiley-Blackwell; 2014

[6] Kalman H. Heritage Planning Principles and Policies. New York, NY: Routledge; 2014

[7] Joe M. Adaptive Reuse and Rehabilitation: Connecting Historic Preservation and Affordable Housing Developments in Seattle. Washington: University of Washington; 2015. Available from: https://digital.lib.washington.edu/researchworks/bitstream/handle/1773/34199/Joe_washington_0250O_14677.pdf?sequence=1

[8] Rowberry R. Alleviating Barcelona's public housing shortages through historic properties. Madrid: Revista de Derecho Urbanístico y Medio Ambiente; pp. 157-176. Available from: https://readingroom.law.gsu.edu/cgi/viewcontent.cgi?referer=https://www.google.ca/&httpsredir=1&article=2999&context=faculty_pub

[9] Li Y, Zhao L, Huang J, Law A. Research frameworks, methodologies, and assessment methods concerning the adaptive reuse of architectural heritage: A review. Built Heritage. 2021;**5**(6):121-132

[10] Mohamed R, Boyle R, Yang AY, Tangari J. Adaptive reuse: A review and analysis of its relationship to the 3 Es of sustainability. Facilities. 2017;**35**(3/4):138-154

[11] Tsenkova S, editor. Historic Places: Adaptive Reuse Strategies for Heritage Conservation. Calgary: Cities, Policy & Planning Research Series. School of Architecture, Planning & Landscape. 2021. Available from: https://sapl.ucalgary.ca/sites/default/files/HistoricPlaces_1.pdf [Accessed: November 6, 2022]

[12] Bullen P, Love P. The rhetoric of adaptive reuse or reality of demolition: Views from the field. Cities. 2010;**27**(4):215-224

[13] Foster G. Circular economy strategies for adaptive reuse of cultural heritage buildings to reduce environmental impacts. Resources, Conservation and Recycling. 2020;**152**:1-14

[14] Meeks S. Place matters—Why reusing old buildings can revitalize your community. Public Management. 2017;**2017**:6-9

[15] Young E, Chan E. Implementation challenges to the adaptive reuse of heritage buildings: Towards the goals of sustainable, low carbon cities. Habitat International. 2012;**36**(3):352-361

[16] Tsenkova S. Energy Efficiency and Affordable Housing—Design, Policy and

Practice for the Social Housing Sector of Canadian Cities. Heidelberg: Springer Nature Switzerland AG; 2021b

[17] Legacy C, Davison G, Liu E. Delivering social housing: Examining the nexus between social housing and democratic planning. Housing, Theory and Society. 2016;**33**(3):324-341

[18] Joseph M, Khare E, editors. What Works to Promote Inclusive, Equitable Mixed-Income Communities. Cleveland: National Institute for Mixed Income Communities; 2020

[19] Shipley R, Utz S, Parsons M. Does adaptive reuse pay? A study of the business of building renovation in Ontario, Canada. International Journal of Heritage Studies. 2006;**12**(6):505-520

[20] Tsenkova S, Witwer M. Bridging the gap: Policy instruments to encourage private sector provision of affordable rental housing in Alberta. Canadian Journal of Urban Research. 2011;**20**(1):52-80

[21] Province of Saskatchewan. Seniors to Gain More Affordable Housing in Regina Downtown. 2005. Available from: Province of Saskatchewan: http://www.saskatchewan.ca/government/news-and-media/2005/april/25/seniors-to-gain-more-affordable-housing-in-regina-downtown [Accessed: October 26, 2018]

[22] Leslieville Historical Society. Coxwell Stables. 2020. Acailable from: https://leslievillehistory.com/coxwell-stables/ [Accessed: October 7, 2021]

[23] Novak C. Living small: The prospects for tiny houses as a solution for affordable housing in Vancouver. 2016. Available from: https://www.chelseanovak.ca/ living-small/living-small-the-prospects-for-tiny-houses-as-a-solution-for-affordable-housing-invancouver/4/

[24] Gilderbloom J, Hanka MJ, Ambrosius JD. Historic preservation's impact on job creation, property values, and environmental sustainability. Journal of Urbanism: International Research on Placemaking and Urban Sustainability. 2009;**2**(2):83-101. DOI: 10.1080/17549170903056821

[25] Plevoets B, Van Cleempoel K. Adaptive reuse as a strategy towards conservation of cultural heritage: A literature review. In: Brebbia C, Binda L, editors. Structural Repairs and Maintenance of Heritage Architecture XII. Southampton: WIT Press; 2011. pp. 155-164

[26] Kovacs JF, Galvin KJ, Shipley R. Assessing the success of heritage conservation districts: Insights from Ontario, Canada. Cities: The International Journal ofUrban Planning and Policy. 2015;**45**:123-132

[27] Brooker G, Stone S. Rereadings 2: Interior Architecture and the Design Principles of Remodelling Existing Buildings. London: RIBA Publishing; 2018

[28] Canada P. Standards and Guidelines for the Conservation of Historic Places in Canada. 2nd ed. Ottawa, ON: Parks Canada; 2010

[29] Comité logement Rosemont. Développement de l'îlot central du Technopôle Angus. Montreal: Office de consultation publique de Montreal; 2017. Available from: http://ocpm.qc.ca/sites/ocpm.qc.ca/files/pdf/P88/8.9_comite_logement_rosemont.pdf. [Accessed: November 2, 2018]

[30] Canada P. State of Canada's Natural and Cultural Heritage Places - 2021 Report. Ottawa, ON: Parks Canada; 2021. Available from: https://parks.canada.ca/agence-agency/bib-lib/rapports-reports/elnhc-scnhp/2021 [Accessed: July 4, 2022]

Section 4

Mitigating Social Inequality

Chapter 8

Contemporary Challenges and Future Strategies to Mitigate Social Inequality in Urban Housing: An Austrian Perspective

Andreas Koch

Abstract

Urban housing in Europe is increasingly challenged by decreasing affordability and availability for an economically and demographically diversified population. Furthermore, residents become more dependent on housing policies that primarily satisfy (global) market demands instead of social needs. This chapter addresses these challenges by focusing on the economic (commodification), spatial (territorialization) and political (centralization) domains of housing. Based on a critical discussion of these domains, the chapter then presents strategies that are supposed to help mitigate social inequality in housing markets. While adaptations of the legal framework contribute to strengthening the social functions of housing, applying relational geography help release the municipals' highly competitive hunting for residents within a containerized imagination of local planning. The promotion of commoning practices considers the need for neighborhood engagement to articulate concerns of the local communities. Communalism is vital, and the chapter proposes strategies to achieve this political state in neighborhood communities. This plea is illustrated by taking an Austrian perspective on housing policy.

Keywords: decommodification, deterritorialization, decentralization, communalism, social infrastructure, relational geography, commoning

1. Introduction

The current housing economy and policy in market-based countries such as Austria are characterized by interrelated structural, functional and processual facts. One of these facts is a prevailing territorial planning paradigm, which rests primarily upon an internal dominance of local residential development and largely ignores the regional scale of a joint housing development. Furthermore, municipalities with their containerized spatial thinking of land use policies produce a situation of competition among each other to gain (high-income) residents and take less care on low(er)-income households.

A second fact is given with a predominant focus on the capital value of a housing unit and consequently de-valuing the use value of a home as a place of shelter and social interaction. That is, building housing units obeys primarily market rules, with problematic side-effects, among others are housing vacancy, touristic use of dwellings or temporarily used second homes. In other words, commodification outperforms societal needs.

Closely interrelated with commodification and territorialization is, thirdly, the centralization of housing and neighborhood policy situated in the municipal government. Even though policy decisions are legitimated by representative-democratic principles, the local government largely ignores the participation and involvement of neighborhoods affected by housing planning decisions.

A socially more sustainable future for housing has to deal with these politically challenging facts, which jeopardize local social cohesion and affect poverty and inequality negatively. Approaches discussed in this chapter focus on these three domains and consider transformations in, firstly, the legal framework, and secondly, in a changing understanding of the geographical framing of planning by introducing a relational spatial model of local and trans-local housing development. Thirdly, a strong plea for commoning practices is outlined, allowing for a local-democratic allocation of housing and neighborhood relations based on societal needs instead of market-driven mechanisms. All these approaches are supposed to contribute to housing poverty and inequality mitigation. They will be discussed in an Austrian urban context, excluding Vienna due to its exceptional social housing policy. Vienna's housing policy is characterized by a significantly higher proportion of social housing than any other Austrian city. This policy dates back to the post-First-World-War period when the city started to build dwellings in public municipal ownership (Gemeindebau). In addition, social housing companies have contributed to this housing segment for decades, which means that today 45% of Vienna's housing market is government-sponsored [1]. **Figure 1** illustrates the legal status of housing at the level of federal states.

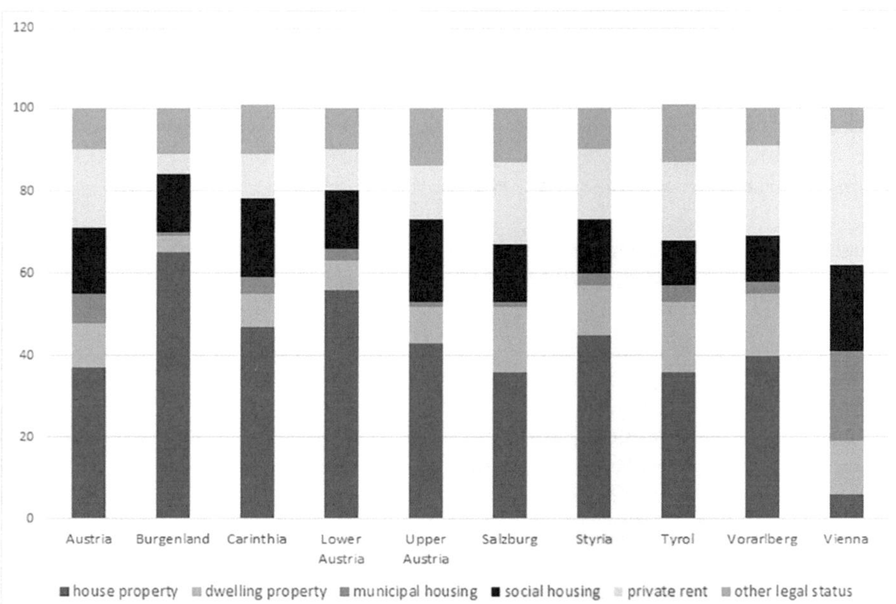

Figure 1.
Legal status of housing per federal state (in percent). Source: ([2], p. 29).

2. Contemporary challenges in urban housing

Present urban housing problems and conflicts result, to a large part, in the prevailing neoliberal market regime, which privileges capital value against use value, promotes competition among cities on residents and values the national or global reputation of real estate markets higher than local residents' needs. The following sections shed some light on these conflicts by taking three perspectives: the economic perspective of commodification, the geographical perspective of territorialization and the political perspective of centralization.

2.1 Commodification of housing

Current housing policy in market economies is impacted by a charged relationship in three interrelated domains. While the structural domain is struggling with the home as a market object versus the home as an object of social inclusion and participation, the functional domain is rivaling between the home as a strategy of capital accumulation and the home as a place to perform various activities such as interacting with others, working, recreating and learning. The material domain, at last, competes with the home as a commodity on the one hand and the home as a private place of shelter and safe haven of personal identity preservation on the other. These three domains are not static, neither historically nor geographically. In fact, they depend on individual needs and socio-cultural habits. Variations in income correlate with biographical life cycles and household composition, which, in turn, often affect residential migration patterns with shifting housing expectations. In addition, an increasing number of single-person households – of younger and older persons – is to be taken into account. New forms of cohabitation, beyond the yet dominating two-generational family, are increasing. However, anything but commonly offered since "[a]lternative forms of housing such as social co-housing, or housing associations […] are growing but still marginal compared to mainstream architecture" ([3], p. 76 f.).

Due to the prevailing belief in the neoliberal paradigm of markets, this charged relationship is significantly imbalanced toward market competition, capital accumulation and commodification. Commodification is defined as the transformation of a 'thing' – a good, a service or even the own body – into a tradable unit on markets, followed by the subjugation of use value to exchange or capital value [4]. The problem of commodification of housing units rests upon weakening its meaning as social infrastructure and promoting capital profit orientation instead. Critical urban geography and policy stress the fundamental failure of housing markets to adequately address the social or community relevance of housing since markets rely on supply and demand rather than needs [5]. The commodification of housing is fueled by accumulated capital searching for long-term and secure profitable reinvestments [6, 7].

The growing albeit contested profit-seeking housing commodification regime relies on a legal framework at the Austrian national and EU supranational scale that prioritizes private property against collective property [8] and competition against public services with only a few exceptions [9]. Public housing assistance is strictly defined and dedicated to unemployed persons or those who work in low-income sectors. It must not affect the private housing market negatively regarding rents and return. Austrian public housing regulations are comparatively comprehensive, notwithstanding, seeking to reach low- and middle-income households. In addition to the formal legal framing of private property, the legal coding of capital became an important step to safeguard the aim of accumulation against other interests. This fact has thoroughly been outlined by Pistor [10].

The consequences of housing commodification are far-reaching and encompass housing vacancy, a commercially touristic utilization of homes, and the spread of second homes. In the city of Salzburg, for example, about 4–5%of all housing units are estimated vacant (not occupied for one year at least) [11]. According to a recent study in Salzburg [12], the touristic use of homes is another push factor that withdraws apartments from the housing market. The Airbnb platform offers 700 apartments, equivalent to almost 1% of the existing housing stock. The number of urban second homes, utilized only for some weeks during the year (however, including a small proportion of apartments occupied by students), is significant in larger Austrian cities. The last national census of 2011 reveals 18.3% in Innsbruck, 17.1% in Salzburg, 15.8% in Graz and 13.7% in Linz [12].

These developments indicate the growing problem of safeguarding the social infrastructure nature of housing and call for political interventions as outlined by the United Nations 'Sustainable Development Goals (SDGs)' or the 'The Right to Adequate Housing'. Both declarations are dedicated to harnessing market-driven profit-making by increasing awareness of all human beings' basic needs, as expressed by the 'leaving no one behind' proclamation. The Right to Adequate Housing principles encompasses, among other things, the "right to choose one's residence, to determine where to live and to freedom of movement", "equal and non-discriminatory access to adequate housing" or "security of tenure" [13]. SDG 11 pursues the goal of safe and affordable housing by 2030 [14]. These goals most explicitly consider the circumstances of the poor and economically deprived.

2.2 Territorialization of spatial organization

From a functional and demand-side perspective, housing can be conceived as living at a particular place at a particular time. The living place is one node of a network of locations where other activities such as working, shopping for food or other things, learning, attending cultural events or doing sports are performed. These functional-spatial networks are not topologically fixed in size, scale and scope but may change with varying working places, biographical requirements, household composition or personal needs and aspirations. However, they remain relational in their spatial structure, and the use value of a dwelling depends critically on the respective topology.

From a market and supply-side perspective, housing relies crucially on a territorial, containerized spatial organization. The monetary assessment of a particular housing unit may be partly influenced by quality standards of the building itself and interior appointments but is determined mainly by the location. The location, in turn, is evaluated comparatively by the value of the neighboring properties. As comparison is a core requirement in market economies to strive for reinvesting accumulated capital, the containerized space serves this need perfectly. Thus, the exchange value of a dwelling depends critically on the respective territoriality.

Although the binary between territorial and topological organizations of political, economic and social functions represents empirical facts in a too simplistic manner, its core idea can claim some justification. While globalization and supra-national political institutionalizations established spaces of flows – of goods, capital, information and people – and networks of global cities [15, 16], developments such as regulating (refugee) migration, combatting pandemic or coding capital are instances of attempts of re-territorialization. However, any attempts to conflate territorial with topological geographies must be wary of fetishizing either of these spaces.

A predominantly territorial, containerized imagination of spatial organization nurtures competition among cities and municipalities in attracting (high-income) households because their public budgets rely strongly on taxes redistributed on the number of primary residents. Demographically shrinking municipalities thus lose income and attention, and they are simultaneously enforced to preserve the still existing infrastructure from dereliction. Demographically growing municipalities must locate new housing blocks and infrastructure, and will yield new income by reallocating taxes after some time. Due to the high attractiveness of most Austrian large cities, suburban municipalities benefit from this competition of people. This is, however, not always and necessarily the case, as medium-sized cities like Eisenerz in Styria illustrate. Eisenerz, a town of 3800 inhabitants in 2020, lost more than 9100 people over the past 70 years. On the other hand, the population of some neighboring municipalities likes, for example, Trofaiach grew remarkably during the same period, partly from households that moved from Eisenerz [17].

The competition between municipalities is geographically volatile since real estate investment strategies have to cope with changing households' lifestyle preferences [18], transforming regional economic and labor markets, and staging mainstream aspirations of where to live the most satisfying life [19]. The main phases had been urbanization, suburbanization, counter-urbanization, and re-urbanization. Gentrification is perhaps the most prominent phenomenon of the current re-urbanization of middle- and upper-class households [20, 21]. Due to an accelerating number of heat days and tropical nights or other climate-change effects [22], and because of experienced challenges of corona measures, a renaissance of suburbanization might likely happen in the near future.

A containerized spatial organization of housing (markets) relates closely to the problem of "methodological nationalism", which equates society with the nation state's territory. The methodology construes internal homogeneity and external exclusion to secure own political and economic interests. This principle, in other words, legitimates global inequality [23]. Its principle idea can be scaled down to the local level and applied methodologically to the administrative units of municipalities. Complementing this unilateral orientation with a relational spatial organization would mitigate housing poverty because it relieves the comparative competition between territories.

2.3 Centralization of political decisions

Spatial planning in Austria – like in other European federal states – is organized hierarchically, with a strong position of local municipalities developing and executing plans that transform agriculturally used land into plots dedicated to housing, public or commercial use. However, the local territorial planning authority is simultaneously embedded into a system of the regional, federal state and national planning levels, which delineate their targets for future spatial development according to explicit efforts of economic, social and ecological sustainability and efficiency. While the latter levels are meant to be recommendations to be considered compulsory by the local level, the local planning manifestations rest upon national laws, i.e., decisions of the local government must be executed on its territory.

Political processes of land allocation and dedication of agricultural land, forests or brownfield zones to be transformed into plots for commercial or residential buildings are, by and large, limited to the local government and subsequent public administration. Public participation is highly formalized by raising objections against already existing plans. A proactive inclusion of engaged citizens from the beginning of the planning process, which would have the power to intervene through local-democratic procedures, is not given in general. Due to the far-reaching and comprehensive legal protection of plots and built housing units by private property laws, citizens in general and affected neighbors, in particular, have thus no direct intervention capabilities. However, local (urban) planning rules set the frame on the type and degree of building and land use to which property owners have to obey.

Besides the hardly existing public involvement in the local spatial development of neighborhoods, a public discussion about a neighborhood's architecture is likewise not envisaged. Usually, it is performed as an interplay between architects and officially designated experts, representing the local government's opinion [24].

A problematic implication of this authoritarian top-down-driven spatial figuration of our housing (and living) environment is an epic standardization of housing design, following a conservative role model of the two-generational family. As a result, each housing unit has its predefined function with little flexibility for multi-purpose activities throughout the day or week. Such a standardized architecture largely ignores the increase of different living forms and household communities that evolved over the past three or so decades. Furthermore, home-based COVID-19 measures have uncovered the weakness of missing multi-functional places for working, learning, eating and sleeping.

Another problem with this planning policy is a lack of control of the local government. Although elected representatives decide how planning principles are to be interpreted in a particular case, it nevertheless often follows political pressure, dictated by allegedly economic constraints of maximizing profits. A lack of control reveals a positive feedback loop of capital accumulation aspirations, territorial competition and local selfishness, supported by a vertical and horizontal planning structure. In other words, commodification overrules social infrastructure, territorial boundedness overrules community networks and statist planning bureaucracy overrules inclusive local participation and civic engagement. The next chapter will discuss approaches that cope with these critical structures, functions and processes.

3. Future housing strategies to mitigate social inequality

Distinguishing the three domains of "commodification", "territorialization" and "centralization" is one option to cope with housing policies in neoliberal market economies. It offers a distinctive variation of views by considering economic, geographical and political perspectives on possible future housing developments. We will start with the economic domain of commodification and discuss some opportunities to de-commodify housing. By adding relational and commoning strategies to disqualify unilateral politics of territorialization and centralization, a re-enforcing process between the three "de" counter-domains – de-commodification, deterritorialization and decentralization – is supposed to be taken place.

3.1 Adapting the legal framework

Housing commodification can be mitigated through a deliberate reform of the legal planning framework without getting inevitably in conflict with the overarching private property legislation. One option in this regard is to reactivate housing vacancies (including, where possible, office vacancies). Vacancy rates are significant in most Austrian cities, such as Salzburg, as has been mentioned above. The federal states of Salzburg and Tyrol have adopted a vacancy levy in July 2022, whereby the fee depends on the size and the construction type. For instance, in Salzburg, a maximum of 400 € (and 800 € for newly built apartments) can be charged annually by the municipality for housing units up to 40 square meters. The overall maximum charge is 2500 € (5000 € respectively) [25]. Several exceptions exist, some of which are reasonable from the de-commodification standpoint: derelict apartments are excluded, as are those with care-dependent persons who moved to sheltered accommodation. Other exceptions like housing vacancies dedicated to the future retirement of the next generation are less well justified.

A second and closely related option would be a revision of the second-home status. A strict(er) definition of "second homes" is needed and should be reduced to only a few exceptions for students or persons who live in the city for at least six months (verification can be obtained through power meter measurements).

The growing touristic usage of apartments is another profit-seeking market development that fuels commodification since turnover is significantly higher than it, is with permanent residents. Public control of the proper use of apartments is difficult to realize because landlords and landladies can command their private property as they wish to do. However, political intervention in the social housing sectors is possible as they have dedicated rules on how to use the homes accordingly. Austrian cities have recently – and partly successfully – started campaigns to persecute and penalize commercial offers of social housing units, for example, at Airbnb.

Another major driving force of housing commodification, in general, is not caused by the housing unit itself but the land underneath. If farmland is planned to be transformed into residential land use by the municipal government, then the capital value is commonly skyrocketing, most likely in urban and suburban regions. With this declarative shift of land use, the landowners become very rich. Although they are legally required to enforce the governmental decision – it is, in fact, based on an application submitted by them – within a timeframe of 10 years, very often nothing happens except that the capital value of the land is further growing. A de-commodification policy on dedicated residential (or commercial) land use would be to (i) reduce the 10-year timeframe significantly, (ii) execute the law consequently and (iii) introduce a tax on the profits generated with the land use change.

A further political instrument in housing de-commodification would be the introduction of temporarily limited and regionally adapted building moratoria. The idea is to pause the construction of new residential complexes in the future and thus interrupt the profit dynamics. Moratoria have been valued as a proper strategy in general [26] and have been introduced recently in some Austrian municipalities, for example, "Haus im Ennstal" and "Velden am Wörthersee" [27].

In order to strengthen collaboration between municipalities in housing planning, the local planning rules should be legally and strictly tied to the regional level. This shift would mitigate both the competition for residents and companies and the lavish commerce with land and resources. While the planning of housing estates would leave the territorially bounded container space, the concrete realization of where to build new apartments would remain within the local territory. This option, however, would affect a transformation of spatial thinking (and public budgeting) and will, therefore, be discussed in the next chapter.

3.2 Applying relational geographies

A second domain appropriate to transform contemporary housing market structures and functions rests upon surmounting the unilateral fixation on the territorial geography of market organization. This perspective can claim some evidence since "[n]eoliberalisation and the opening up of global markets, as well as intensifying the modes and level of exploitation that take place through capital accumulation, has been a major driving force in the disruption of territory as a factor in political economy, culture and identity" ([28], p. 1646).

The idea of surmounting, however, does not imply a substitution of territorial thinking with topological spatialities but is meant to conflate both in an emancipatory understanding of local housing needs. In so doing, we, according to Ince [29], refer to an "anarchist approach to territory that foregrounds bordering as a legitimate spatial strategy that refuses and moves beyond a statist-capitalist framework for understanding the role and nature of territorial practices, and that can produce emancipatory spaces in the process".

Relational conceptions of space can claim their particular success because they deconcentrate (spatial relations) and decentralize (power relations) to the concrete local spots, aiming – at least implicitly – to oppose authoritarian constraints. By pursuing a relational strategy, territorial space properties do not disappear but diminish their exclusionary characteristics. Role models of this spatial type have been available for a long time, such as consortia between research and education institutions or global city networks [30], to mention just two examples.

When reflecting on the meaning of relational spaces, we must keep in mind that a municipality's spatial gestalt is arbitrary due to different reasons. Any ideas and visions about housing architecture and neighborhood infrastructure are restricted to plans, programs and political decisions, which must be applied to the entire territory in equal measure. For example, car parking capacity rules – underground and ground-level – of larger housing complexes in Austrian cities are to be followed by housing construction companies with little flexibility concerning their number and distribution. It is, however, not possible today to develop a housing complex without car parking slots, even though the new residents would prefer such an option because the complex is well connected to public transportation from their point of view.

Therefore, a neighborhood community should be politically able to decide on their local circumstances of housing and environment by interlinking territorial and relational geographies. This instance can be generalized to housing architecture (co-housing), public neighborhood space design (potential conflicts between younger and older residents), social mixing (private property and social housing) and functional mixing (housing with labor, recreation, and shops). This way, territorial practices are re-introduced to contribute to solving future housing needs decentrally.

In addition, a relational perspective includes novel approaches in municipal collaboration, albeit not necessarily between territorially bounded units. Future housing questions are then discussed between several places by considering social and ecological concerns that interrelate local with regional requirements. A prerequisite for its achievement would be a transformation of the current budgetary fundament a municipality depends on, which is income and commercial tax redistribution to a large degree. Examples of collaborative school planning and national park planning do exist but are not mandatory and committed. Similarly, the central place approach in settlement planning follows relational principles at a regional level.

The network geography would complement territorial planning politics by emphasizing local housing needs in relation to other – already existing or yet developing – structural links (infrastructure) that connect functional activities in an appropriate manner, whereby appropriateness is defined by the local collectives that belong to the networks at hand. It would help to reduce the competitive-selfish momentum of exclusionary territorial planning due to reciprocal mechanisms of exchanges of money and tangible and intangible infrastructures [31]. Furthermore, it would help to promote local community participation and engagement with their natural and social environments. This idea refers to communalism or libertarian socialism and seeks to conflate individual autonomy with collective commitment [32]. As a local-democratic initiative, it is seen as an essential ingredient to conquering housing commodification and an authoritarian housing policy and philosophy.

3.3 Promoting Commoning practices

The third domain, coping with decentralization and relocation strategies, refers to commons and commoning. Commons can be defined as arrangements of co-production, co-consumption and co-utilization of goods, services, land and, not least, housing. They rearrange access to all these 'things' through collective structures of participation in political decision-making processes at the local scale [33]. The idea of commons also includes alternative forms of cohabitation by promoting the sharing of devices, offering rooms for common use or the co-caring for children and older people.

Thanks to an increased awareness of the far-reaching adverse effects of neoliberal capitalism, commons received a renaissance over the last two or so decades. Their aspirations are committed to all domains of the sustainability and socio-ecological transformation discourses since commons do not only strive for intermingling individual wellbeing with collective wellbeing, which tightens relationships among residents. Anarchistic-inspired commons also take ecological concerns seriously into account, as they explicitly attempt to reduce resource and energy consumption by substituting market-individualistic with community-individualistic strategies of housing and mobility patterns. Furthermore, those commons establish plans and programs that combine private with collective property, promote sharing economy models and prioritize public transportation offers.

A recently published "micro-political manifesto" [34] to housing cooperatives proposes several political activities to approach a revised understanding of housing as a core part of social infrastructure. Among other things, the manifesto sticks up for

- activating micro-political actors and networks
- developing skills for commoning practice
- claiming housing as a shared resource
- claiming responsibility through collective ownership

- incorporating collective land use within planning frameworks
- establishing cooperative properties as social and cultural assets ([34], p. 11–39).

In order to realize a transformation toward commoning practices in the fields of housing and neighborhood relations, different tactics need to be applied, considering local circumstances explicitly. In the context of a cooperative housing project in Sao Paolo, Brazil, Ventura [35] stresses a strong interlinkage between local (neighborhood) and global (the city) housing policies – "that micro-political organization can only unfold its full potential if it also gains macro-political relevance" to oppose to prevailing neoliberal capitalism of housing economy. Similarly, Woldeyessus [36] focuses on strategies that bridge "representational collectives" (government) and "collective representations" (governance), i.e. strategies that mutually interlink local parliaments of elected delegates with collectives representing civil society movements.

Another example that promotes affordable non-profit housing and a transformation from property to belonging is the German Mietshäuser Syndikat [37]. The legal organization rests upon the rule to prioritize the "utility value" against its "capital value". This rule is achieved by a "model of divided ownership: every housing project is owned by a limited liability company (Haus GmbH), which – in turn – is owned by two shareholders, the Hausverein (Dwellers' Association) and the Mietshäuser Syndikat. [...] While the structure of the limited liability company allows the Hausverein to be self-sufficient in all questions concerning the use and management of the house, the Mietshäuser Syndikat has the mandate to safeguard the ownership status of a building and to prevent the property from privatization" ([37], p. 136). Another model for financing collective housing property is micro-funding, which has been applied more often to the so-called Global South [38].

The outlined commoning approaches inspired us to reflect on potential translations into the Austrian context. Any attempts to decentralize future housing and neighborhood policies would increase their success if micro-political activities were interrelated with macro-political framing programs. These programs would pave the way for a growing diversity – in number and nature – of housing forms to oppose the highly standardized housing architecture, following the idea of a mono-functional use of rooms and praising the two-generational family as a role model.

Explorable references in a European context are, for example, the pioneering housing and neighborhood design activities in Zurich, Switzerland (Kraftwerk, Kalkbreite, Mehr als Wohnen) [39]. Intellectually grounded on libertarian communalism and outlined in programmatic writings such as Bookchin [40] and P.M. [41], their practical transformations reveal versatile ideas and imaginations about alternative ways of cohabitation. "This utopian attempt articulated for the first time the idea of agency through design, shattering the existing urban order and replacing its conventional typologies with a free-flowing, autonomous entity in collective ownership" ([39], p. 182).

To mitigate poverty and reduce social inequality, increased diversity of housing architectures, allowing for multi-functional uses, is needed. This need is also reflected in a changing composition of households, with an increased number of single-person and single-parent as well as multi-person homes beyond the kinship model. Moreover, new migration and mobility patterns require novel approaches to diversified housing supplies, ranging from tiny mobile homes and temporary housing communities to fully-furnished apartments offered by local communities or municipalities. Unfortunately, the housing market supply has not yet sufficiently anticipated these new demand patterns. However, the first experiments of social co-housing with privately and collectively used rooms, car-free neighborhoods and collectively shared goods and services exist in Austria [42]. A prominent example is given with the habiTAT [43]. An overview of alternative, participatory housing models can be found in [44, 45] for Austria, and [46] for Switzerland.

The idea of communitarian housing can be extended to the co-utilization of goods and services needed in a neighborhood community. Bikes and cars, tools, garden devices, books, kitchens and many more things are worth being included as part of a local sharing economy. Furthermore, the re-invention of subsistence framing as an innovative economic approach in the food sector to provide households within a regional range would also be part of the commoning vision. Formally situated between small-scale urban gardening and commercial urban farming, subsistence farming increases neighborhood food autonomy and decreases transportation efforts effectively and sustainably. While urban gardening is very popular in European cities [47], subsistence farming is still in its infancy. Further models of subsistence could be used in the energy sector. The production of local solar, water or wind power likewise contributes to enhancing independence from national or global energy production – a fact whose significance will rise in the future due to climate change policy efforts and the revealed severe dependency on energy resources from the Russian government and other authoritarian states.

One recently realized example of a communitarian housing project is "Cooperative Housing Volkersdorf" near Graz [48]. This project offers 28 apartments for 63 people to date. Several of the characteristics highlighted above have been realized in Volkersdorf, such as a partial subsistence economy in food production, renewable energy production, use of ecological resources in housing construction and co-working opportunities nearby.

The success of community-driven housing can be enhanced if intermediary nodes between local communities and urban housing representatives are introduced. Those nodes – commonly referred to as neighborhood managers – serve as bridging relations between the macro- and micro-levels in urban politics on the one hand and relations among residents of the cooperative(s) on the other. This relational capacity can be defined as "performed urbanism", which differs from patterns of "self-organized urbanism" (the idealistic form of solving conflicts) and "instrumentalized urbanism" (exploiting urban amenities with no or little personal contribution) [49].

Models of performed urbanism have been established in newly created larger housing complexes in, for example, Salzburg, for some years. They complement other social services at the city district level. Neighborhood management is organized jointly by the city administration and social associations, like the Diakoniewerk [50]. The neighborhood managers provide services to satisfy the residents' heterogeneous needs and contribute to moderating controversy. Also, public rooms can be used for different events that bring residents closer together. This approach of vertically and horizontally intermediary collaboration can serve as a role model for self-organized housing cooperatives in one way or the other.

4. Conclusion

The three "de"-perspectives culminate in a common political center-point that we can adequately circumscribe with the notions of "communalism" or "libertarian municipalism", two concepts that have been coined by Murray Bookchin [40]. Among other things, they express the necessity to withdraw particular goods and services from the market because they satisfy the basic needs of all people, irrespective of income, social status or any other discriminatory difference. They are common goods. Besides goods and services of health and education, mobility, food and water, safe and livable environment, it is housing in its material and social-relational conditions. Libertarian municipalism conflates the singularity of women and men with their socially derived individual aspirations on the one hand and the collectivity of the local neighborhood in which they are socially embedded on the other.

In order to re-strengthen the commitment to the local spatial and social environment, political decentralization becomes one core ingredient in future housing policies. To put the housing's function of social infrastructure to the fore and defend its primary purpose as a means of cohabitation aligns with the UN's Sustainable Development Goals and does not inevitably jeopardize its value-preserving meaning. In fact, this latter meaning remains relevant, considering the ecological sustainability of our societal transformation in the Anthropocene. The commodification of the home is neither necessary nor desirable any longer, which, in turn, fosters decommodification as another core ingredient in future housing policy. The local is not an isolated and unconnected spot on the earth's surface like a sand grain on a sand hill, but a node in a network of interrelated nodes. The relations may functionally vary, but the essential point is that distance is not metric but relational. This functional shift produces a different pattern of collaboration and cohabitation, surmounting the traditional understanding of territorial inclusion and exclusion. Not least, deterritorialization turns out to be the third core ingredient in future housing policy. There is no ideal or perfect point where all three "de"-perspectives merge. The decisive moment is that they have left the corner points of their counterparts.

Acknowledgements

The publication has been funded by the University of Salzburg's Open Access Funds.

Author details

Andreas Koch
University of Salzburg, Salzburg, Austria

*Address all correspondence to: andreas.koch@plus.ac.at

References

[1] Stadt Wien. Hintergrund: Gemeindebau und Sozialer Wohnbau in Wien. Available from: https://www.wien.gv.at/presse/hintergrund/wohnbau. [Accessed: August 9, 2022]

[2] Statistik Austria. Wohnen. Zahlen, Daten und Indikatoren der Wohnstatistik. 2022. Available from: https://www.statistik.at/fileadmin/publications/Wohnen-2021.pdf. [Accessed: August 9, 2022]

[3] Koch A. The impact of contemporary housing functions on its social sustainability. In: Almusaed A, Almssad A, editors. Sustainable Housing. London: IntechOpen; 2022. pp. 75-90. DOI: 10.5772/intechopen.99277

[4] Hermann C. The Critique of Commodification: Contours of a Post-Capitalist Society. Oxford: Oxford University Press; 2021. DOI: 10.1093/oso/9780197576755.001.0001. p.218

[5] Holm A. Wiederkehr der Wohnungsfrage. In: Bundeszentrale für politische Bildung, editor. Gesucht! Gefunden? Alte und neue Wohnungsfragen. Bonn: Selbstverlag; 2019. pp. 98-111

[6] Harvey D. Seventeen Contradictions and the End of Capitalism. London: Profile Books; 2014. p. 338

[7] Harvey D. The urban process under capitalism: A framework for analysis. In: Fyfe NR, Kenny JT, editors. The Urban Geography Reader. London and New York: Routledge; 2005. pp. 109-120

[8] Pechmann AV. Die Eigentumsfrage im 21. Jahrhundert. Bielefeld: transcript Verlag; 2021. p. 228

[9] Shah N. Europäische Union und der soziale Wohnbau – ein Spannungsverhältnis?! In: Club of Vienna, editor. Brennpunkt Wohnbau. Wien, Berlin: Mandelbaum; 2019. pp. 9-28

[10] Pistor K. Code of Capital: How the Law Creates Wealth and Inequality. Princeton: Princeton University Press; 2019. p. 297

[11] SIR Salzburger Institut für Raumordnung und Wohnen. Wohnungsleerstand in der Stadt Salzburg. 2015. Available from: https://www.salzburg.gv.at/bauenwohnen_/Documents/endbericht_wohnungsleerstand_final.pdf. [Accessed: July 26, 2022]

[12] Van-Hametner A, Smigiel C, Kautzschmann K, Zeller C. Die Wohnungsfrage abseits der Metropolen: Wohnen in Salzburg zwischen touristischer Nachfrage und Finanzanlagen. Geographica Helvetica. 2019;**74**:235-248. DOI: 10.5194/gh-74-235-2019

[13] UN Habitat. The Right to Adequate Housing. Fact Sheet No. 21. 2015. Available from: https://unhabitat.org/the-right-to-adequate-housing-fact-sheet-no-21rev-1. [Accessed: July 26, 2022]

[14] UN United Nations. Sustainable Development Goals. 2022. Available from: https://sdgs.un.org/goals. [Accessed: July 26, 2022]

[15] Appadurai A. The Social Life of Things: Commodities in Cultural Perspective. Cambridge: Cambridge University Press; 1988. p. 348

[16] Castells M. The Rise of the Network Society: The Information Age: Economy, Society, and Culture. 2nd ed. Vol. I. Chichester: Wiley; 2010. p. 656

[17] Gstach I, Kapferer E, Koch A, Sedmak C. Sozialatlas Steirische Eisenstraße. Wien: Mandelbaum; 2013. p. 179

[18] Waddell PA. Household Choice and Urban Structure. Aldershot: Ashgate; 1997. p. 125

[19] Florida R. The Rise of the Creative Class. New York: Basic Books; 2002. p. 512

[20] Smith N. Gentrification, the frontier, and the restructuring of urban space. In: Fyfe NR, Kenny JT, editors. The Urban Geography Reader. New York: Routledge; 2005. pp. 128-137

[21] Vollmer L. Strategien gegen Gentrifizierung. Stuttgart: Schmetterling Verlag; 2018. p. 163

[22] Laue F. Heat-stress related climate-change adaptation in informal urban communities. In: Ley A, Ashiq Ur Rahman A, Fokdal J, editors. Housing and Human Settlements in a World of Change. Bielefeld: transcript Verlag; 2020. pp. 189-218. DOI: 10.14361/9783839449424

[23] Beck U, Einleitung PA. In: Beck U, Poferl A, editors. Große Armut, großer Reichtum. Zur Transnationalisierung sozialer Ungleichheit. Berlin: Suhrkamp Verlag; 2010. pp. 9-22

[24] Heindl G. Stadtkonflikte. Wien: Mandelbaum Verlag; 2020. p. 270

[25] Salzburg24.at. Land Salzburg beschließt Leerstandsabgabe. Available from: https://www.salzburg24.at/news/salzburg/land-salzburg-beschliesst-leerstandsabgabe-123831163. [Accessed: August 8, 2022]

[26] Fuhrhop D. Verbietet das Bauen! Munich: oekom; 2015. p. 189

[27] Brunner S. Schluss mit dem Ausverkauf. Die Zeit 2021. 2021. Available from: https://www.zeit.de/2021/35/tourismusregionen-oesterreich-widerstand-investoren-zweitwohnsitz. [Accessed: July 8, 2022]

[28] Flint C. Political geography I: Globalization, metapolitical geographies and everyday life. Progress in Human Geography. 2002;**26**(3):391-400

[29] Ince A. In the Shell of the old: Anarchist geographies of Territorialisation. Antipode. 2012;**44**(5):1645-1666. DOI: 10.1111/j.1467-8330.2012.01029.x

[30] Barber B. If Mayors Ruled the World. New Haven & London: Yale University Press; 2014. p. 416

[31] Kapferer E, Koch A, Sedmak C. Strengthening Intangible Infrastructures: A Commitment to Communities. Newcastle upon Tyne: Cambridge Scholar Publishing; 2013. p. 323

[32] Bookchin M. Die nächste Revolution. Libertärer Kommunalismus und die Zukunft der Linken. 2nd ed. Münster: Unrast Verlag; 2020. p. 220

[33] Ostrom E. Governing the Commons: The Evolution of Institutions for Collective Action. Cambridge: Cambridge University Press; 2015. p. 296

[34] Delz S, Hehl R, Ventura P. Housing the Co-op. A Micro-Political Manifesto. Berlin: Ruby Press; 2020. p. 240

[35] Ventura P. The Mutirao program in Brazil: Co-operative housing production

intersecting micro- and macro-political spheres. In: Delz S, Hehl R, Ventura P, editors. Housing the Co-op. A Micro-Political Manifesto. Berlin: Ruby Press; 2020. pp. 79-92

[36] Woldeyessus B. Co-operative Co-production: The case of post-occupancy housing Co-operatives at Gotera site in Addis Ababa, Ethiopia. In: Delz S, Hehl R, Ventura P, editors. Housing the Co-op. A Micro-Political Manifesto. Berlin: Ruby Press; 2020. pp. 69-77

[37] Heilgemeir A. The Mietshäuser Syndikat: A structural answer to anti-speculative and self-governed housing production. In: Delz S, Hehl R, Ventura P, editors. Housing the Co-op. A Micro-Political Manifesto. Berlin: Ruby Press; 2020. pp. 135-147

[38] Delz S. Co-operative capital: Making use of collective micro-fund models for housing production. In: Delz S, Hehl R, Ventura P, editors. Housing the Co-op. A Micro-Political Manifesto. Berlin: Ruby Press; 2020. pp. 121-132

[39] Davidovici I. Co-op City: Utopia and pragmatism in Zurich's Co-operative housing. In: Delz S, Hehl R, Ventura P, editors. Housing the Co-op. A Micro-Political Manifesto. Berlin: Ruby Press; 2020. pp. 177-192

[40] Bookchin M. From Urbanization to Cities. The Politics of Democratic Municipalism. Chico, USA, Edinburgh UK: AK Press; 2021. p. 315

[41] P.M. bolo'bolo. Zurich: Paranoia City Verlag; 2015. p. 231

[42] Jany A. Experiment Wohnbau. Die partizipative Architektur des Modell Steiermark. Berlin: Jovis Verlag; 2019. p. 200

[43] Mietshäuser Syndikat Österreich. 2022. Available from: https://www.syndikat.org/en/. [Accessed: August 8, 2022]

[44] Weber M, Amann W, Mundt A. Praxisempfehlungen zu gemeinschaftsorientiertem Wohnen. Projektbericht (Neunkirchen/Wien, im Auftrag des Landes Niederösterreich). 2019. Available from: https://docplayer.org/175813282-Praxisempfehlungen-zu-gemeinschaftsorientiertem-wohnen-projektbericht.html. [Accessed: September 8, 2022]

[45] Gemeinsamwohnen.at. 2022. Available from: http://www.gemeinsamwohnen.at/links. [Accessed: September 8, 2022]

[46] Eine SS. Geschichte des gemeinschaftlichen Wohnens. Modelle des urbanen Zusammenlebens. Basel: Birkhäuser; 2019. p. 324

[47] Biedermann A, Ripperger A-L. Urban Gardening und Stadtentwicklung. Wiesbaden: Springer Spektrum; 2017. p. 236. DOI: 10.1007/978-3-658-18698-2

[48] Die Wogen. Wohnprojekte-Genossenschaft. KooWo Volkersdorf – Nähe Graz. 2022. Available from: https://diewogen.at/unsere-projekte/koowo-volkersdorf/. [Accessed: September 8, 2022]

[49] Smigiel C, Koch A. Formen von Urbanität in der kapitalistischen Moderne und ihre Effekte. Polylog. 2018;**39**:19-33

[50] GBV Gemeinnützige Bauvereinigungen. "Wohnen im Alter" – Fachtagung der Gemeinnützigen Bauvereinigungen Salzburg. Available from: https://www.wohnen.net/de/presse/detail.asp?id=133&tit=„Wohnen%2520im%2520Alter“%2520–%2520Fachtagung%2520der%2520Gemeinnuetzigen%2520Bauvereinigungen%2520Salzburgs. [Accessed: September 8, 2022]

Chapter 9

Perspective Chapter: Is Expecting Older People to Downsize to Help Solve the Nation's Housing Crisis Really Such a Good Move?

Tony Watts

Abstract

On a regular basis the cry goes up: "Older people should downsize to free up their houses for younger families." However, it's not quite as straightforward as it's made out to be, maintains older people's campaigner Tony Watts OBE. Moreover, the lack of a suitable housing supply in the right locations means that significant opportunities to improve the health and wellbeing of older people... and reduce public expenditure on health and care... are being missed.

Keywords: later life housing, retirement living, downsizing, age-friendly communities, housing crisis

1. Introduction

A key highlight of the UK 2021 Census [1] was the inexorable ageing of our national population: the "baby boom" that followed the Second World War, combined with increasing longevity, mean that almost one in five of us is now aged 65 and over. In some parts, such as Norfolk, it is one in three. Those figures are set to grow still further in the years ahead.

The implications – economically, socially and politically – are huge. And one of the biggest of these is where (and how) we house our ever-ageing population, many of whom will have restricted mobility or specific care or support needs.

2. The pros and cons of downsizing

In the latter part of this paper I look at how a more considered and co-ordinated approach to meeting the housing needs of older people would actually represent a national opportunity: the key to better care, fewer (and shorter) hospital stays and a more integrated society. But first I'd like to address the related and highly controversial issue of downsizing that, at the moment, is diverting attention away from seizing this opportunity.

It's not unusual in our society for older people to be portrayed either as an economic and social burden and a roadblock to progress, and/or as the owners of a disproportionate amount of the nation's wealth and assets... with housing the prime example of that.

Blaming older people for the country's housing crisis is not only discriminatory but also demonstrably untrue: the problem is far more deep-rooted and borne of decades of undersupply against a background of a rising population. Despite that, older people are routinely characterised as selfishly "rattling around" in the large houses they were fortunate enough to purchase before housing became unaffordable to a large swathe of the population... and they are regularly being exhorted to make way for others. In November 2021 [2], then Housing Minister Chris Pincher became one of the latest, telling a House of Lords committee that almost four in ten properties were currently "under-occupied" and could be better used by younger families with children.

He went on to insist that the Government was keen to encourage housebuilders to create more developments suitable for pensioners. Having heard a string of Ministers make similar noises over the years, but seen little to further the process (not least, failing to change obstructive planning regimes) you'll excuse me if I do not hold my breath.

In fact, research by WPI Economics and Homes for Later Living [3] estimates that three million older people would like to downsize, yet only about 7000 new retirement properties are built each year. Their "key national policy recommendation is for a 10-year programme of building 30,000 retirement properties every year", which would represent a complete step change in the planning and development process.

For those seeking smaller, more manageable and accessible homes, bungalows have long been hugely popular and could easily form an important part of this equation... if enough were being built. But in 2020, just 1942 came out of the ground, according to the National House Building Council: around 2% of total housing supply [4], against a figure of 26,000 in 1987. Further research in 2021 by McCarthy Stone [5] found that 70% of over-65 s would consider moving to a bungalow. Older people cannot be asked to move to housing that is not there.

Agreed, on the face of it, downsizing to more suitable housing makes good sense on all sides. The prospect of reducing maintenance and heating costs, along with lower Council Tax payments for those downsizing... perhaps even releasing some much-needed capital along the way; plus a chance to get families into larger accommodation. Moreover, two million homes occupied by older people are currently deemed "not decent", invariably because they cannot afford to adapt or improve them; reducing that figure by enabling older singles and couples to move into housing that was in good repair, warm and accessible would also be very welcome.

Further, if a proportion of those new retirement units were to be located on our currently decaying High Streets, they could play a big role in revitalising our town and city centres, as has been set out by organisations including ARCO (Associated Retirement Community Operators) [6] – as well as enable the occupants to live close to the shops and other services they need in their daily lives.

3. Obstacles to opportunity

And, yes, dedicated retirement developments are being built, but nowhere near enough (or in the right places) to meaningfully move any dial, or for the calls for more older people to downsize to represent a realistic possibility. Only 2.5% of the UK's 29 million dwellings are technically defined as "retirement housing" according to a recent report from the Urban Land Institute [7] while the number of purpose-built homes also offering care services is far less, at around 0.7% of UK housing stock.

Neither are the statistics likely to significantly improve any time soon. According to Laing & Buisson in their 2021 review of the senior housing sector [8], "We expect the number of specialist seniors housing units in the UK will grow by 9% over the next five years to just short of 820,000 units. Yet, even with this forecast expansion, the rate of delivery will still be dwarfed by the UK's ageing population, deepening the existing imbalance between supply and demand."

They go on to say: "The benefits of a larger seniors housing sector to society are substantial. Yet, just 20% of local authorities in England have supportive planning policies or sites allocated to seniors housing."

In their July 2022 report, "Later Living: The £30bn Baby Boomer Challenge" [9], property experts JLL estimate that "an ageing population will require an additional 75,000 later living homes and 30,000 care homes over the next four years." Neither of which are remotely likely to happen.

Critics of retirement housing proposals during planning applications regularly point to existing schemes not being fully occupied. But this can be down to a series of factors, not least location: many prospective older home buyers will resist moving away from their immediate neighbourhood – with its ready-made network of support and social contact. To square that particular circle, we need more choice local to where people currently live, and with so few units being built each year, that level of choice currently does not exist.

Small wonder there is a lot of resentment building up among those who are being made to feel fingers are being wagged in their direction: yes, there's a housing crisis, but whose fault is that? The plain fact is, we simply aren't building enough new homes full stop – let alone enough suitable housing for older people, many of whom would elect to live in mainstream housing if it was better adapted to their needs.

Lest we forget, our population continues to grow: albeit Brexit has skewed figures (probably) short term, it's around 0.5% per annum – an additional 350,000 people a year. Against that, in England we reached the dizzy heights of 160,000 new homes in 2019 before sliding back to 123,000 in 2020 [10]. A 2020 Building Research Establishment (BRE) report found we now have the oldest housing stock in Europe [11] and that, to replace it at current rates, each new build will need to last 1000 years. No wonder we have over four million poor quality homes in the UK, over half of which are occupied by older people.

4. The impacts of inaction

We all know the impacts of the continuing failure to meet housing targets being felt across the whole of the population: spiralling house prices, unaffordable private rents, lengthening waiting lists for social housing along with millions of unhealthy and/or unsafe homes.

But, critically, neither are most of the houses we ARE building fit for purpose... once you take into account the age of the people who are, or will, be living in them. By 2066, a further 8.6 million UK residents will be aged 65 years and over (an additional cohort the size of London [12]). And the fastest increase will be among those aged 85 years and over.

If retirees are to be persuaded into mainstream homes where they can age in place, there's a long way to go. For instance, around 40% of those in the UK aged 60 to 74 report having a lifetime limiting illness, rising to almost 60% at 75 and over [13] ...but DCLG itself estimates [14] a supply gap of up to 45,000 units a year of housing suitable for people with varied support needs, a figure that is growing as each year passes.

And here's the thing: a baby born in 2020 has a 54.3% chance of reaching the age of 90 [15]. You would think it would make sense to ensure that every new home being built would be suitable for everyone to age in place. However, by the time today's young generation reach retirement age, say, or start to live with a disability that makes a mainstream house difficult, there will be even fewer suitable homes available than now.

From its analysis of 324 local housing plans, Habinteg Housing [16] has found that there will just be one new accessible home erected in the next 10 years for every 77 people in the population – down from one for every 67 two years ago.

5. Health implications

So how can meeting the housing needs of our ageing population be turned from a challenge into an opportunity?

The pressure now on individuals, local authorities and health services is to keep people in their own homes for as long as possible rather than enter the care system... but inadequate suitable housing militates against that. If these individuals cannot be supported to live safely in their own homes, they will have to go into care. And while some may be able to pay for themselves, a large number will not... placing the burden on our already-overstretched local authorities.

I view social care, health and housing as three legs of a stool, each of which needs to be in place (and equal in length) for the stool to remain upright. The solution is a coordinated, dual approach: more specialist housing with care and support plugged in, together with making our mainstream housing fit for purpose for the needs of an ageing population. It will require time, commitment and a relatively modest investment, but it will not just be this generation that will benefit, but every generation to follow.

However, given the opportunity to put in place basic "lifetime standards" as a pre-requisite for all new housing, so occupiers could age in place close to their existing support network, Government has opted instead to bow to market pressure and leave it to individual authorities to decide whether developers need to make that small additional investment based on how this would affect a scheme's viability [17].

There are bright spots. The London Plan for instance requires local authorities to ensure that 90% of homes meet "Category 2" standard for accessibility [18] – broadly equivalent to the old "Lifetime Homes" one. But outside of London, less than a quarter of new homes are required to meet the standard. London are also showing the way on wheelchair accessibility, requiring 10% of new housing to meet that higher standard. The Liverpool Local Plan (adopted January 2022) also includes these 90% and 10% figures. If those local authorities can see the benefits of these requirements, why cannot others?

The Government is currently "consulting" on the subject [19], and making some positive noises; we can but hope... but we all know what happens to consultations – especially if there are vested interests to overcome – and not helped by regular ministerial changes at the top. Like others who believe in accessibility standards being made mandatory rather than voluntary, the Housing Champions group to which I belong have made their feelings clear in the consultation process.

What are campaigners like myself arguing for? Well not that much really. Category 2 requires a home to be accessible to most people and fit for purpose for older people, those with reduced mobility and some wheelchair users. The estimated additional cost to developers would be an extra £1400 per home [20]. However, not making them readily accessible will incur a cost we will all pay… this and future generations.

To me, this all comes down to looking at the impact of inappropriate or poor housing on health in a far more integrated, holistic way and breaking down the silos that currently exist between the various departments and organisations with a stake in this, allied with a willingness (or otherwise) to invest in the future.

Another example of this blinkered vision that has thwarted the development of more housing for older people is the regular resistance seen at planning committees throughout the land… resistance from local residents who insist that starter homes for young people should take priority, but also resistance from councillors who fear the impact of having more older people to care for in their community – ignoring the fact that these are the very older people currently living locally in unsuitable homes and whose care bills will ultimately be paid for by that same local authority.

Agreed, starter, affordable and family-sized homes are needed too. But consider this: every time an older person moves out of a home that has become too big for them and into a purpose-built unit, it frees up a home for a younger couple or family to occupy… a virtuous circle. The older person, meanwhile, could have delayed moving into a care setting as well as freed up capital that will help them fund that care. It should not always be "either-or", as I'll come onto later.

6. The financial benefits to society of investing in housing

Looking at the bigger picture, there is also a direct correlation between poorly maintained or designed housing and ill health and mortalities. According to the BRE, the estimated costs of poor housing to the NHS is £1.4 billion a year [21]. Even if the country reached the Government target of 300,000 new homes each year [22] and built them to Category 2 standards the total cost of that would be £420 million… just one third of the NHS figure.

Invest in older people's housing and the whole of society benefits, but that opportunity is currently being ignored.

According to The Strategic Society [23], aggregated savings to the State for each new specialist retirement housing unit built amounted to £83,100 over a period of 10 years – once you take into account reduced health and care needs, a reduction in local authority entitlements and the benefit of first-time buyers not having to rent and so receiving housing benefits. Equally, how much is it currently costing the nation for some 530,000 long-term sick and disabled over 65-year-olds to be living in a non-decent home [24]? The current energy crisis is highlighting the problem that so many elderly people have in keeping their home warm and dry when it is poorly insulated, draughty or subject to damp.

The failure to adequately adapt and repair the existing homes of older people represents a huge missed opportunity to improve the health and wellbeing of a large swathe of the population and reduce public expenditure in the process. In the Centre for Ageing Better's 2021 report, "Home & Dry, the need for decent homes in later life" [25], compiled in association with Care & Repair England, the conclusion is drawn that: "After decades of improvements and year-on-year decreases in the number of non-decent homes, the rate of improvement is stalling for all ages. In the case of households headed by someone aged over 75 years old, the trend has actually reversed and the number of non-decent homes has risen from 533,000 in 2012 to 701,000 in 2017. This age group are also disproportionately likely to be living in a non-decent home compared to younger age groups (more than 1 in 5 of over 75-year-old households)."

Further, they report that: "New analysis for this report shows that the NHS spends £513 million alone on first year treatment costs for over 55s living in the poorest quality housing. An investment of £4.3 billion could repair all these homes – a cost that would be paid back in just over eight years, and would immeasurably improve the quality of life for millions of people, now and in the future."

The lack of relatively small amounts of funding is cited as the primary obstacle to dealing with this issue: "Around one third of all non-decent homes could be repaired for £1,000. Previous funding to address housing disrepair, such as means-tested grants for lower income homeowners, has been withdrawn in recent years. In 1983-84 the national allocated funding for private-sector housing improvement and repair was £1.1 billion. By 2010-11 this was down to £317 million, and then ceased completely the following financial year."

As ever, it's all a question of priorities... and votes. As of May 2022 [26], the Government had provided over £22 billion in equity loans for 355,634 starter homes since 2013, creating lots of nice headlines for an investment of just over £60,000 per home. But the UK needs "Later Homes" too. Where is the support to make that happen?

If you enable older people to remain safe and independent in their own home, and so age in place, many will be able to rely on their existing informal social and familial networks for basic support... which not only reduces the burdens on health and social care but also delays their need to go into specialist care. Surely it's time to make this a national priority, benefiting all generations in the process?

But some 95% of older people live in accommodation lacking even basic accessibility features [27], so when mobility or health issues strike, many find it hard to move about their own homes, leaving them to live in just one or two rooms. A percentage will inevitably suffer a fall as a direct result of living in a house that has poor accessibility or is not fitted with even basic mobility equipment such as grab rails and ramps.

Why is this factor so important? According to NICE [28]: "Falls are *the leading cause of mortality resulting from injury in people aged 75 and older in the UK. Around 30% of adults who are over 65 and living at home will experience at least one fall a year (approximately 2.5 million people in England). Falls and fractures in people aged 65 and over account for over four million hospital bed days each year in England alone.*" [29]

Moreover, getting those people back home again swiftly, and "unblocking" the bed they are occupying, can be a major problem if the assessment undertaken before they can be discharged finds their home is unsuitable for them to return safely.

7. Who will pay?

According to the Chartered Institute of Housing, "For many older people, being able to live independently at home, or in a homely setting, for as long as possible can be hugely beneficial for both physical and mental health, and local connections to family, friends and community can help to reduce social isolation and loneliness.

This can also reduce the need for more costly interventions such as a stay in hospital or a move to residential care" [30].

But who should foot the bill for this?

There is, of course, the perfectly sensible argument that many of those in houses that need to be repaired or adapted could very well afford to pay for the work. Agreed, some can, but not all... 67% of those currently living in "pensioner poverty" in the UK are owner-occupiers [31]. They are "asset rich and cash poor": owning their own home (which in many cases has spiralled in value since they purchased it) but struggling to afford to heat, repair, adapt and maintain it.

Making access to equity release type loans more flexible and competitive could be one part of the solution, allowing older homeowners to release and live on part of the value of their home. Another would be to adequately fund "handy man" care and repair services that can carry out the small but vital jobs that make homes safer and more accessible: sadly, many councils have seen this this as one area where economies can be made.

Care & Repair England research shows that 300 people can be helped by a handyperson for the same cost as one place in a care home for a year [32]. This really exemplifies the principle of "investing to save" as many of those whose move into care can be delayed will rely on some measure of local authority funding. It is perhaps symbolic that Care & Repair England itself recently closed (April 2022) after 36 years because it could no longer secure funding for its work. As it said in its statement on closure: "Despite the strong evidence and data about the connections between ageing, disadvantage, poor housing and health, effective policies, funding and action to tackle non-decent homes are all notable by their absence" [33].

Alongside that, providing better mechanisms for the 76% of older people (aged 55 and over) who own their own homes [34] to move to more appropriate housing could enable more to fund their own future. But why not accelerate that process and incentivise older people in homes that are too large for them to downsize? Better still, provide funds that would enable retirement living developers to incorporate more affordable homes in their projects... something that is also a big barrier for many.

8. Older people's perspectives

What, then, do older people themselves think about this? Prescribing solutions will never work without their buy-in. Quite simply, if older people had more say in the decision-making processes that affect their lives, many of the issues I've covered here could be resolved.

I am one of the one of Older People's Housing Champions appointed by Care & Repair England, a national network of older activists and sector specialists who have for the last decade or so worked to improve housing and related services for our ageing population – primarily by acting as sounding boards for public and private sector bodies including national Government.

One of our tasks has been to generate a "manifesto" [35] of the actions we believe would make the biggest contribution to ensuring older people can live independently as long as possible in homes that are safe, warm, comfortable and accessible. Because so many people live (and continue to live) in mainstream housing, part of our focus is on adapting and maintaining existing housing stock; but we also have strong views on future housing too. These are the key tenets of that manifesto:

1. Home adaptations can extend safe, independent living at home. They also have a significant impact on health and wellbeing, by preventing falls and accidents. Home adaptations assistance should be mandatory, delivered quickly, efficiently and be a core part of future integrated health, social care and housing systems.

2. A significant number of older people, particularly low-income homeowners, live in poor quality, cold homes which have a negative impact on their health. Many homes occupied by older people require repairs and maintenance. There should be nationwide provision of practical, affordable housing repair and adaptation services – including home improvement agencies and handyperson services for older people in all housing sectors.

3. Small "healthy at home" grants or low-cost loans for essential repairs and improvements (including heating systems) should be made available for disadvantaged older people resulting in benefits for individuals and society.

4. Local authority house condition audits should be re-introduced, alongside private sector housing renewal programmes, to tackle disrepair and prevent existing housing stock decline.

5. All new homes should be built to accessible standards and be suitable for further adaptation.

6. Build more innovative mainstream housing of a design and size that is particularly suitable for later life – e.g. using HAPPI 24, DWELL25 principles.

7. Build a wider range of specialist and supported housing for those with later life care and support needs.

8. Demand for specialist housing is likely to outstrip supply. More supported housing options for older people of all income groups are needed – both to rent and to buy.

9. A national source of independent and specialist housing, care and finance information, combined with impartial local one-to-one advice and support, is urgently needed for older people, their carers and professionals. This would include a register of accessible, adaptable housing to help people locate suitable homes when their needs change.

10. Older people – experts by experience – need to be engaged and involved in developing and delivering housing solutions and strategies for later life at all levels locally and nationally.

Older people, then, know what they want... but their needs are not currently being addressed. Why is this latter point so important? Because unless you create housing that older people want to live in, they will not move there. Yet still you see developers applying for consents with no consultations with local people on what they might want to buy or rent. The result is all too often rejection at the planning stage or developments struggling to fill up.

The Localism Act means that developers and planners are obligated to take on board local views. The principles of developing any new product for any marketplace entails market testing, but this does not always seem to take place with new housing for older people: would it not make sense for a developer to fully engage with a community before the pre-app stage and determine precisely what is really needed and wanted locally? As well as securing a speedier consent by demonstrating demand and heading off criticisms from local nimbys, they would also be able to refine their offer at an earlier (and less expensive) stage, then go on to sell or rent their apartments or houses far more readily.

I will add another key factor here: developers should create product that can be seen as aspirational, not a last option. Many of us go through our lives looking to progress up the housing ladder and find it problematic to make a move to a smaller house if it is perceived as a "downward" move. Not everyone wants to move into a shoebox or lose their garden; not everyone wants to move away from their community; not everyone wants to live well away from shops and services or somewhere that has no public transport links.

Provide more choice, listen to what local older people feel would meet their specific needs, change the concept of "downsizing" to "rightsizing", a repositioning that I have argued for over many years now, and you might encourage more to make what (for many) will be their last move.

9. Conclusions

In essence, if society is to meet the challenges of housing an ageing population, concerted action needs to be taken that breaks down the silo thinking that currently dogs the planning process, and instead involves all of the stakeholders: local authorities, health trusts, developers, the Government and older people themselves.

Alongside more resources being dedicated to adapt and repair the mainstream housing that many older people currently occupy, the end goal should be a greater choice of more dedicated and/or suitable housing enabling people to age in place – not just in larger developments but in much smaller ones too, that allow more of them to be pepper-potted into our towns, cities and villages and encourage the occupiers to engage with their local communities... a mix of generations living together and supporting each other.

Many older people really do not want to live in what are often labelled as "old people's ghettoes"... not least because most of us have a problem with seeing ourselves as being as old as we actually are! Moreover, the sights and sounds of children playing and the interaction with younger generations is something many of us would miss.

If there is a more holistic approach to add to the mix that would meet the challenges and needs of an ever-ageing society, instead of homogenous developments squeezed into parcels of land aimed at just one part of the market, why not go for a far more mixed approach, and engage with local people to co-design new self-sustaining communities? As I suggested earlier, this overcomes the "either-or" decision we so often see at planning.

These would be integrated and intergenerational developments, with each generation looking out for other, and with the potential for people to transition to larger/smaller/differently designed houses as their needs changed. It would not cost any more to build, but should cut the cost of care for individuals and the community... and enable people to age in place close to an informal support network.

Of course, that's not a new idea at all. For centuries, most of us lived in just such communities. We called them villages, and they worked pretty well. Perhaps we need to rebadge them as something whizzier like "smart villages" and the idea might regain traction...

Author details

Tony Watts
Director of EngAgeNet and Older People's Housing Champion, Bristol, England

*Address all correspondence to: tony@hartleywatts.co.uk

References

[1] Census 2021. Available from: https://census.gov.uk/

[2] Meeting the UK's Housing Demand. Available from: https://committees.parliament.uk/work/1328/meeting-the-uks-housing-demand/publications/

[3] Homes for Later Living/Silver Saviours for the High Street. Available from: https://homesforlaterliving.org/wp-content/uploads/2021/02/Homes-For-Later-Living-Silver-saviours-for-the-high-street-.pdf

[4] https://www.nhbc.co.uk/binaries/content/assets/nhbc/media-centre/stats/2020-new-home-statistics-review.pdf

[5] https://www.mccarthyandstone.co.uk/media-centre/national-press/uks-over-65s-demand-bungalow-renaissance-after-plummeting-build-rates/

[6] https://www.arcouk.org/blog/reigniting-town-centres-and-high-streets-through-older-peoples-housing

[7] Later Living: Housing with Care Guide. Available from: https://labmonline.co.uk/features/the-urban-land-institute-uli-launches-later-living-housing-with-care-guide/

[8] Seniors Housing Annual Review. Available from: https://content.knightfrank.com/research/1854/documents/en/seniors-housing-annual-performance-review-2021-8481.pdf

[9] Later Living: The 30bn Baby Boomer Challenge. Available from: https://www.jll.co.uk/en/trends-and-insights/research/later-living-the-30bn-baby-boomer-challenge

[10] New Home Statistics Review. 2020. Available from: https://www.nhbc.co.uk/media-centre/statistics/2021/01/29/2020-new-home-statistics-review

[11] The Housing Stock of the United Kingdom. Available from: https://files.bregroup.com/bretrust/The-Housing-Stock-of-the-United-Kingdom_Report_BRE-Trust.pdf

[12] Living Longer: How Our Population is Changing and Why it Matters. Available from: https://www.ons.gov.uk/peoplepopulationandcommunity/birthsdeathsandmarriages/ageing/articles/livinglongerhowourpopulationischangingandwhyitmatters/2018-08-13

[13] Later Life in the United Kingdom 2019. Available from: https://www.ageuk.org.uk/globalassets/age-uk/documents/reports-and-publications/later_life_uk_factsheet.pdf

[14] Housing for an Ageing Population – Planning Ahead. Available from: https://www.gov.uk/government/uploads/system/uploads/attachment_data/file/314647/ukafa-housing-ageing-population.pdf

[15] Mortality in England and Wales: Past and Projected Trends in Average Lifespan. Available from: https://www.ons.gov.uk/peoplepopulationandcommunity/birthsdeathsandmarriages/lifeexpectancies/articles/mortalityinenglandandwales/pastandprojectedtrendsinaveragelifespan

[16] Forecast for Accessible Homes. 2020. Available from: https://www.habinteg.org.uk/localplans/

[17] https://publications.parliament.uk/pa/cm201617/cmselect/cmwomeq/631/63107.htm

[18] London Plan Guidance, Housing Design Standards. Available from: https://www.london.gov.uk/sites/default/files/housing_design_standards_lpg_-_publish_for_consultation_1.pdf

[19] https://www.gov.uk/government/consultations/raising-accessibility-standards-for-new-homes#full-publication-update-history

[20] Raising Accessibility Standards for New Homes. Available from: https://assets.publishing.service.gov.uk/government/uploads/system/uploads/attachment_data/file/930274/200813_con_doc_-_final__1_.pdf

[21] The Cost of Poor Housing in England. Available from: https://bregroup.com/press-releases/bre-report-finds-poor-housing-is-costing-nhs-1-4bn-a-year/

[22] Tackling the Undersupply of Housing in England. Available from: https://commonslibrary.parliament.uk/research-briefings/cbp-7671/

[23] Valuing Retirement Housing. Available from: https://strategicsociety.org.uk/wp-content/uploads/2016/08/Valuing-Retirement-Housing.pdf

[24] Special Analysis of English Housing Survey 11-12 Data undertaken by BRE for Care & Repair England in 2016

[25] Home and Dry: The Need for Decent Homes in Later Life. Available from: https://ageing-better.org.uk/sites/default/files/2021-08/Home-and-dry-decent-homes-later-life.pdf

[26] 350,000 Households Supported to Buy a Home Through Help to Buy. Available from: https://www.gov.uk/government/news/350000-households-supported-to-buy-a-home-through-help-to-buy

[27] Age UK Agenda for Later Life. 2015. Available from: www.ageuk.org.uk/Documents/EN-GB/For-professionals/Policy/agenda_for_later_life_2015_full_report.pdf

[28] Falls in Older People. Available from: https://www.nice.org.uk/guidance/qs86/documents/draft-quality-standard-2

[29] Falls in Older People Quality Standard [QS86]. 2015. Available from: https://www.nice.org.uk/guidance/qs86/chapter/Quality-statements

[30] https://www.cih.org/media/qg4mdmtz/health-and-social-care-consultation-response.pdf

[31] Pannell J, Aldridge H, Kenway P. Market Assessment of Housing Options for Older People. London: New Policy Institute; 2012 Available from: https://www.npi.org.uk/publications/housing-and-homelessness/market-assessment-housing-options-older-people/

[32] Care & Repair England. Ageing Well at Home. England: Nottingham Care & Repair; 2014

[33] https://careandrepair-england.org.uk/news/care-repair-england-closing

[34] Living Linger: Changes on Housing Tenure Over Time. Available from: https://www.ons.gov.uk/peoplepopulationandcommunity/birthsdeathsandmarriages/

ageing/articles/livinglonger/changesinhousingtenureovertime

[35] https://housingactionblog.files.wordpress.com/2021/12/housing-champions-housing-manifesto-final.pdf